Mr. Hume is of opinion that the Alps are part of the high chain of Mountains seen at a distance from the sea coast

Covered with Snow

Impenetrable Brush

AUSTRALIAN ALPS

Twisden R. 60 yds wide

Mangaroo Plains

Fine forest Country

Dampiers Range

High Granite Mountain Timber lofty and good

good forest land

Kanguroos numerous

Kings R.

Ovens R. 80 yds wide

undulating country

Deep rocky and unimpenetrable gullies into a picturesque and extensive valley thinly wooded

Hume R. 100 yds wide rich alluvial soil open country

Bells Valley Mt.

Pudding Stone

Granite

Flat Country thickly timbered and apparently subject to inundation no high land visible

Brown Slate

Iron Bark

Bloomfields Valley

Undulating Country

Longitude East

Extensive and undulating Downs of alternate weed and Plain called by the natives "Tarnoo"

Mt. Dissappointment

Slate and Sandstone

Brisbanes Range

This Range supposed to the Northern and South... wate...

Mt. Byron

High land to the

Mounts
Romulus
Remus

Meehans Peak

apparently fine Country

The Tourists found in their way back that most of the high lands East of the Twisden as far as the Hume terminate abruptly in a flat Country

To the North and North West the horizon is unbroken

114

CAVAN STATION

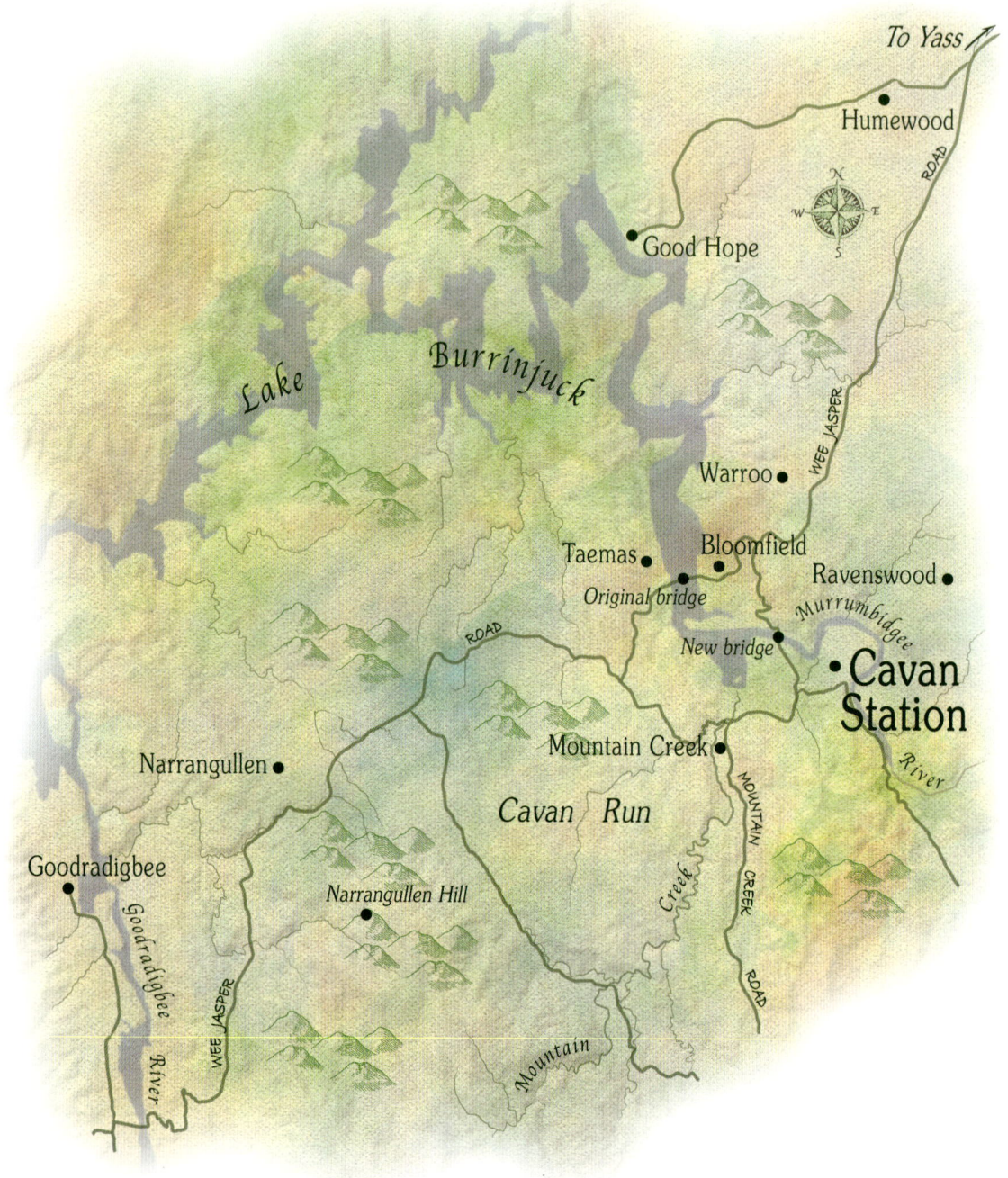

To Yass

Humewood

Good Hope

Lake Burrinjuck

WEE JASPER ROAD

Warroo

Taemas

Bloomfield

Original bridge

Ravenswood

Murrumbidgee

ROAD

New bridge

Cavan
Station

Narrangullen

Mountain Creek

Cavan Run

River

Goodradigbee

Narrangullen Hill

WEE JASPER

Goodradigbee River

Creek

MOUNTAIN CREEK ROAD

Mountain

To dear Lucy and Richard —
with much love
— Nicola xx   March 2019

CAVAN STATION

Its early history, the Riley legacy and the Murdoch vision

BY NICOLA CRICHTON-BROWN

HarperCollins*Publishers*

'Sir, I am a true, simple labourer: I earn what I eat, get what I wear, hate no man, envy no man's happiness, am happy for other men's good fortune and satisfied with my own bad fortune, and the source of my greatest pride is watching my ewes graze and my lambs feed.'

— William Shakespeare, *As You Like It*, Act 3 scene 2

*In memory of Lorna Phillips, who contributed so much to this book
and who sadly passed away before its publication.*

CONTENTS

FOREWORD

CAVAN, EXCEPT AT PEAK TIMES, is only one hour's drive from the nation's capital. It is unknown to most Australians, and yet in the mid-nineteenth century it was one of those big sheep runs which quickly shaped this nation's economic and social history. Wool was then, much more than today, a vital commodity in world commerce and manufacturing, and Australia was top of the wool ladder.

This is the story of Cavan and its geology, plants, livestock and fascinating people.

A crucial episode in its history was when the small sailing ship *Sir George Osborne* crossed the world to Sydney with 198 European sheep, living below deck and fed mainly with hay and gruel. During the voyage of nineteen weeks, sometimes in stormy seas or high temperatures, the young Edward Riley cared for the sheep so capably, day and night, that only four died.

From their first Australian home, in the Hunter Valley in 1825, many descendants of these Saxon sheep a few years later helped to stock the rolling pastures of Cavan, which William Riley bought in July 1834. The first wooden house was erected at Cavan just before Melbourne and Adelaide were born. The first seasons' bales of merino wool were carried in a procession of bullock drays all the way to Sydney Harbour where they were loaded onto England-bound ships.

The Saxon breed of merino had originally grazed not far from the German city of Dresden – so heavily bombed in the Second World War – and were almost treated as pets; they wore rugs in winter and were sheltered in barns and sheds. But at faraway Cavan, they flourished in the open air and were protected by the convict shepherds in the day and by the portable open-air enclosures where they crowded together at night. The improved Australian breed of Saxons produced a larger fleece at shearing time, and its longer and stronger and yet softer fibres had the qualities needed by the new spinning and combing techniques of the industrialised woollen mills in the north of England. Isolated sheep stations like Cavan were part of the global chain of production that made possible the industrial revolution in Europe.

Visitors to Cavan even then marvelled at the beauty of the scenery. Below the little weatherboard homestead flowed the Murrumbidgee River, almost at the start of its long winding journey to the sea near Adelaide; and for much of the year the water glistened in the sunlight, especially when it crossed the rapids and rockbars, though at times it was 'inky black' and too dangerous for swimmers. In the mid distance were rolling grasslands and limestone ridges, while far ahead stood the Snowy Mountains.

The author of this engaging history, Nicola Crichton-Brown, introduces us to the owners, one by one. William Riley, aged only twenty-nine, committed suicide on the property in 1836, just when many of the early obstacles had been overcome. In the 1860s the manager was James Calvert. When aged only nineteen, he had joined the exploring party led by Ludwig Leichhardt, who eventually found a circuitous overland route from the Brisbane district to Port Essington in Arnhem Land: the longest expedition in the continent up to that time. As manager of Cavan, Calvert faced the flood of April 1870, and as the river waters spread, he had a rowing boat tied to a peach tree while all night he relied on the light of a carriage lamp to illuminate the peril he faced.

He had recently married Australia's most prolific female novelist, the native-born Louisa Atkinson, and in articles written just after that frightening flood at Cavan, she described for *The Sydney Morning Herald* the captivating scenery of the sheep station and especially the limestone formations, packed with ancient shellfish fossils and even coral. The fossil beds made the hillsides resemble a series of ribbons, one above the other, and from distant viewpoints the slopes displayed the pattern and colours of a ploughed field. Only a few miles from the Cavan homestead, we learn, is a magnificent collection of embedded fossils, known as 'Shearsby's Wallpaper', and named after a young bank clerk and amateur geologist from the nearby town of Yass.

Once so isolated, Cavan eventually became linked to the swiftly changing world. In 1876, the railway from Sydney reached Yass. A year later, the first post office was opened at Cavan, though letters from Sydney were sometimes stale by the time they arrived.

The owners of Cavan came and went, while some chose to stay despite the challenges of low rainfall or poor wool prices or accumulating debts; the property in 1887 was virtually owned by the Bank of New South Wales. The rabbit invasion, creating havoc, called for the vast expense of constructing miles of wire-netting fences. The depression of the early 1890s and the subsequent Federation drought hurt but did not eliminate most of the small farms near the big sheep station.

'Laddie' Castle-Roche, inheriting Cavan from his father, confronted the First World War and saw his wool made into army uniforms. He experienced the astonishing Murrumbidgee flood of May 1925 when the old homestead was left with only the walls and a few veranda posts, while the newer homestead and its extensive garden and lawns, supervised by his wife Enid and a full-time gardener, was more fortunate. Then came the bushfires of January 1939, which burned 10,500 acres of grassland. Many of the injured sheep had to be shot by horsemen with 'rifles strapped to their saddles'.

It was Laddie's descendants who in 1966 sold the property, much shrunken in area, to Rupert Murdoch who was then living in Canberra and running his young national newspaper *The Australian*. We read how he has enlarged the property almost to its old size, revitalising it and improving the pastures, and multiplying the sheep and the Aberdeen Angus, too. So Cavan today is one of the showplaces and pathfinders for the nation's pastoral industry.

This beautifully illustrated book is really a distinctive version of the history of rural Australia, investigated by the author with hard digging and told with spirit.

Geoffrey Blainey
October 2018

INTRODUCTION

Above — *Rupert Murdoch on the veranda at Cavan homestead, 2017*
Previous pages — *Aerial view of Cavan homestead and garden today*
Pages 4–5 — *The main entrance to Cavan*

THERE ARE FEW OCCASIONS when Rupert Murdoch, media giant and multi-billionaire, becomes sentimental, but one of these is when he talks about the Australian bush and, in particular, his pastoral property, Cavan Station. Amongst the oldest holdings in New South Wales and located just south of Yass with double frontage to the Murrumbidgee River, Cavan now embraces 23,000 acres of undulating limestone country and rich river flats, liberally scattered with eucalypts, wattles and kurrajong trees. The landscape is open and expansive, stretching for miles towards the horizon; but it is also dramatic, with prominent, angular ridges and slate-coloured, rocky outcrops, some of immense proportions. It is a landscape worthy of cinematography but, for Murdoch, it is quite simply his pride and joy and the place he calls home.

We meet on a hot and cloudless summer day in January 2017 on the veranda of the sturdy and unpretentious Cavan homestead. A wisteria sprawls above our heads as we sit talking in capacious wicker chairs. The garden and the country beyond appear immaculate in spite of the heat – the piercing light of the Australian sun highlighting the clean contours of the surrounding hills. Murdoch is dressed as if for a day in the yards, in blue denim jeans and a distinctive royal-blue work shirt, embroidered above the left-hand pocket with the name 'Cavan Station' – obligatory kit for all staff on the property. Although Murdoch is also wearing, in a nod perhaps to New York fashion, black sneakers instead of the customary pair of R.M. Williams boots, he instantly comes across as 'one of the team' in a large operation that is helping to set new benchmarks in the management of Australian rural properties.

It would be easy to be cynical about Murdoch's reasons for acquiring Cavan and his professed affection for the bush; but the latter is undeniably genuine, stemming from a childhood roaming the tawny paddocks of Wantabadgery, the historic 19,000-acre Riverina property near Gundagai, purchased by his father, Keith Murdoch, in 1938. As a boy, the young Rupert and his older sister, Helen, spent the spring and autumn school holidays there, with little or no parental supervision, leading an idyllic, carefree existence, riding ponies, hunting rabbits, swimming in the muddy dams and icy Murrumbidgee River, and generally getting up to all sorts of pranks that most children only read about in adventure stories. From the start, Rupert loved Wantabadgery unequivocally – the spaciousness and grandeur of the landscape, the sense of freedom this inspired, and the whole idea of life on the land which today still gives him unalloyed pleasure.

When Keith Murdoch died suddenly in 1952, Wantabadgery was sold, together with another Murdoch property, Booroomba, effectively bringing an end to the halcyon days in the Australian bush of Rupert's youth. Now in charge of a large newspaper business at the age of only twenty-one and newly graduated from Oxford, Rupert had little time for anything other than work, but his affection for the land remained steadfast and undimmed. A taste of a Riverina childhood had left an indelible impression on the young man who was later to live and work almost exclusively in a metropolitan environment. Like so many Australians of his generation, he felt a connection to the bush which, until the explosion of urbanisation following the Second World War, used to be synonymous with the nation's identity.

In 1965, when Murdoch was thirty-three years old, he was, in his own words, 'rather rash' and launched a new daily paper, *The Australian*, which became the first national newspaper in the country. As it was based in Canberra, Murdoch moved there from Sydney, where he had been living with his first wife, Patricia, and his daughter, Prue. The couple were in the process

— *Cavan homestead, c. 1930*

of divorce and, now that he was confidently established in the newspaper business, Murdoch was able to turn his mind to the purchase of a rural property of his own. Enlisting the help of an old family friend, John McEwen, leader of the then Country Party and deputy prime minister, Murdoch was introduced to a third party with strong connections to the land who told him that Cavan was for sale, 'quietly'.

Contacting the then owner, Bill Castle-Roche, who was the fourth generation of his family to live there, Murdoch arranged to meet with Bill in early 1966. Arriving at the homestead – which at that time had a large and curious portico at the front, built in the style of an ancient Greek temple – Murdoch was

— Looking southwest over Cavan from Ravenswood. Murdoch's favourite view

overwhelmed by the beauty and majesty of the setting. He fell in love at first sight with the sweeping views of the mountains and the Murrumbidgee, the magnificent stand of trees along the river and the striking limestone ridges, rising up from the country below and extending, twisted and contorted, for miles around. Nostalgia got the better of him as memories of the freedom and joyful times experienced at Wantabadgery came flooding back, and he contemplated the possibilities of what life at Cavan could be.

It was an informal meeting with Bill, who exuded great warmth and charm as was his custom, but who – it soon became evident – was selling Cavan with some reluctance. Bill took Murdoch on a cursory inspection of the property, and they then had lunch in the homestead dining room. Murdoch vividly remembers that Bill, who was carving the chicken, somehow lost control of the carving knife and fork – possibly due to having consumed what seemed to be a large tankard of gin – so that the chicken flew off the dining table onto the floor! This amusing incident seems only to have oiled the wheels of their negotiations, and a deal was struck after a second meeting in Canberra. Bill agreed to sell Cavan to Murdoch on condition that Bill and his family could continue living in the homestead for another year whilst they completed their new house at Cavan West. Murdoch was happy to allow this, and he used the simple stone cottage at the back of the homestead at the weekends. He had not yet remarried and was living a more or less bachelor existence, though he had by then already met Anna, who would become his second wife.

Cavan was just under 3,000 acres when Murdoch bought it, but it was perfect for him. The property was in good condition, as Bill had always insisted on improvements of only the best quality; and it was reasonably close to Canberra, about fifty miles away. Murdoch began to spend as much time on the property as his work allowed and felt very much at home there. It became his retreat and, ultimately, a place that Murdoch's extended family could be together and develop its cohesion. As he had correctly anticipated, they grew to love Cavan as he had loved Wantabadgery, and today Murdoch's only regret is that, due to business and family commitments in the U.S., his visits are limited and never enough in terms of frequency or duration. One of his favourite spots to sit is on the veranda where we met for our discussion, listening to the birdlife that he points out is unique to Australia. There are so many different species here – sulphur-crested cockatoos, magpies, rosellas, grass parrots, choughs, whipbirds, galahs, wrens, kingfishers, wedge-tailed eagles, wood duck and, of course, kookaburras, whose raucous courtship from the hilltops at dusk, Murdoch enjoys listening to above all else.

Whilst Cavan is a place of enjoyment for Murdoch and his family, it is also very much a commercial enterprise and

Murdoch was overwhelmed by the beauty and majesty of the setting. He fell in love at first sight with the sweeping views of the mountains and the Murrumbidgee, the magnificent stand of trees along the river and the striking limestone ridges, rising up from the country below and extending, twisted and contorted, for miles around.

—

not merely a wealthy man's hobby. A state-of-the-art modern pastoral property, it nonetheless has many traditional features, including a large workforce and horses (instead of motor or quadbikes) as the main form of work transport. As might be expected from Murdoch, 'a tireless man of vision',[1] Cavan has been transformed into a model property in terms of productivity, sustainability and ecological awareness, with a careful eye kept on managing and improving both the stock and the land for future generations. But Murdoch is disarmingly modest and humble regarding these achievements, much of which he credits to his son-in-law, Alasdair Macleod, who is heavily involved in setting the overall direction of the business.

Importantly, Cavan has been returned to its original purpose as a major wool-growing enterprise after several decades of mixed farming. Whilst a small herd of Angus beef cattle is run alongside the sheep, the core business is merino wool, based on Bogo bloodlines with genetics that hark back to the Saxon merino introduced into New South Wales in the mid to late 1820s. One of the earliest importers of the Saxon sheep was Alexander Riley, whose son, William, developed this fine-woolled breed in New South Wales with spectacular success, grazing the sheep on various Riley landholdings, including Cavan, which was purchased by William in 1834. The Riley legacy of transforming the Australian wool industry is widely recognised; what is less known is the role that Cavan played both at the beginning of this occurrence and subsequently. Cavan's history is rich, not only in terms of innovation in wool growing, but more generally in its connection to some of Australia's leading explorers, scientists and writers of the nineteenth century. Its current ownership by Australia's most high-profile family adds to this fascinating and significant past; the Murdoch contribution to the furtherance of excellence in wool growing has become increasingly evident but is, as yet, unsung.

Nicola Crichton-Brown
January 2017

Cavan has been transformed into a model property in terms of productivity, sustainability and ecological awareness, with a careful eye kept on managing and improving both the stock and the land for future generations.

—

Right — *Cavan homestead and front paddock*
Opposite — *Cavan homestead today*

CHAPTER ONE

CAVAN'S ANCIENT PAST

THE UNIQUE GEOLOGY AND PALAEONTOLOGY OF CAVAN

WHILST CAVAN IS KNOWN TODAY as an established grazing property, the land itself at one time lay below the sea and was subjected to enormous seismic activity.[1] Every visitor to Cavan remarks upon the unusual landscape, especially the 'ribbons' of limestone that festoon the hills, which from a distance appear to the onlooker like a ploughed field or, where the ribbons are more widely spaced and broader, like terraces on a Mediterranean hillside. Many of these ridges, however, are sharp and narrow and only a foot or so high, but their leaden colouring nevertheless stands out amid the pasture. In other places, the limestone appears in a large mass and is much more prominent, such as the conspicuous limestone bluff, Clear Hill, south of Cavan homestead; or the huge folds in the landscape of the famous anticline at Taemas, just northwest of Cavan. The latter massive rock formation, now known locally as 'the Shark's Mouth', represents some of the oldest limestone deposits exposed along the Murrumbidgee and reveals the turbulent history of Australia's geological past.

Rising to an altitude of about 3,500 feet, the limestone hills around Cavan have been a source of fascination to geologists and other scientists for almost two centuries, not only due to their idiosyncratic formations, but also because of their extraordinary fossil deposits. One of Cavan's more prominent former residents, the naturalist and author Louisa Atkinson, wrote about the rocks and fossils at Cavan in two articles for *The Sydney Morning Herald* in 1870.[2] She was a friend of the Reverend William Branwhite Clarke, now regarded as the 'Father of Australian Geology', who had begun collecting fossils in the area as early as 1848[3] with the aim of working out the geological age of the local

Above top — *Limestone ridges clearly visible along the Murrumbidgee*
Above — *The Shark's Mouth on Taemas*
Opposite — *Limestone ridges on Cavan, looking east*
Previous pages — *Shearsby's Wallpaper, an example of highly fossiliferous limestone near Cavan*

Every visitor to Cavan remarks upon the unusual landscape, especially the 'ribbons' of limestone that festoon the hills.

—

limestone. No doubt Louisa was aware of his research, and the two may have shared information about their respective finds. Even before this, however, the abundant fossils of corals and marine shellfish had attracted the notice of the first European explorers to the region. Surveyor General Thomas Mitchell collected fossils along the Murrumbidgee in the Yass district on his Australia Felix expedition of 1836,[4] and Count Paul Strzelecki recorded fossils from the nearby Yass Plains in his book, *The Physical Description of New South Wales and Van Diemen's Land* in 1845.

The area would become a tourist attraction as well as a site for scientific discovery. Robert Etheridge, eminent geologist and palaeontologist, wrote in 1889: 'The Murrumbidgee limestone is ... crammed with fossils, especially corals. As a display of these beautiful organisms in natural section, I have never seen its equal.'[5] Whilst there are literally hundreds of places in the Cavan area where fossils can or have been found, Etheridge may well have been thinking of the now much visited heritage-listed site called 'Shearsby's Wallpaper', only a few miles west of the Cavan homestead along the old Taemas Road beside the Murrumbidgee. This exposed section along the river of highly fossiliferous limestone was christened after Alfred Shearsby, the most famous amateur geologist in the region who was widely acknowledged in New South Wales to be an expert in this field.

ALFRED SHEARSBY

Shearsby came to live in Yass in 1898 at the age of twenty-six as an employee of the Australian Joint Stock Bank.[6] His clerical skills were, however, but a microscopic example of his prodigious talents. Despite never having had any tertiary education, he became an amateur geologist and palaeontologist, and a diligent collector of rocks and fossils, but also a biologist and entomologist, as well as an artist and photographer. Apparently inspired by attending a geology students' field trip in the Yass district in 1900 led by a renowned Professor of Geology at Sydney University, Edgeworth David, he may also have had some rudimentary instruction in geology at Fort Street Model School in Sydney, which he attended between 1887 and 1890.

After meeting David, Shearsby immersed himself in geology, which became a lifelong passion. He began writing about his geological discoveries in the Yass district in various scientific journals, describing the surrounding countryside as a 'veritable geological and palaeontological paradise for the collector'.[7] He also started to accompany university professors and lecturers on geological excursions in the district, facilitating access for these groups to various locations where he and others had made important discoveries. He was a significant donor of specimens to a wide variety of museums at home and abroad, and assisted numerous professional and amateur collectors. In recognition of Shearsby's contribution to the understanding of the remarkable geology of the Yass district, especially the Silurian and Devonian limestones, a scholarship in his name was founded at Sydney University by fellow geologists, three years before his death in 1962.

THE DEVONIAN PERIOD (OR THE AGE OF FISHES) IN THE CAVAN DISTRICT

'Shearsby's Wallpaper' is world famous for its assemblage of vast numbers of shells of 'brachiopods', the most common shellfish of the geological period known as Devonian. Sedimentary and volcanic rocks of Devonian age (between 360 million to 420 million years old) occupy almost the entire belt of country between the Murrumbidgee and Goodradigbee Rivers where the original Cavan grant, the adjoining Cavan run and the neighbouring property, Narrangullen, are situated.[8] In that period, Australia formed the eastern part of the vast supercontinent of Gondwana, and as the sea level fluctuated, it inundated what is now the east coast of Australia. As this land mass was closer to the equator than today, the climate was warm and tropical; these two factors combined to encourage the formation of coral beds on the sea floor, close to the land that is now New South Wales and especially in the region around Cavan. Although another supercontinent existed, Euramerica, lying in the northern hemisphere also near the equator, eighty-five percent of the world was nonetheless covered in ocean. With such a large aquatic environment, a huge variety and abundance of fish were spawned across the Devonian globe, giving the period its popular name of the 'Age of Fishes'. Importantly, this was the first time in the history of the planet that vertebrate, or back-boned, animals became numerous.

The area surrounding Cavan is, however, unique in this era. Here, it appears, swam a greater variety of vertebrates in the form of fish than anywhere else recorded in the Devonian world, including a group of strange, heavily armour-plated creatures known as placoderms.[9] These primitive jawed fishes dominated both the shallow tropical seas of the Devonian period, and also the freshwater environments. Most were small or moderate in size, deriving their name from the thick, interlocking bones in the skin that shielded both head and thorax. Some placoderms (e.g. the arthrodires, with the skull and thoracic armour articulated by a well-developed neck joint) were specialised for feeding with tooth-like denticles or shearing blades in their jaws, becoming the main predators of the Devonian seas.

Examples of placoderms from the Cavan area include Buchanosteus, which had smaller denticulate toothplates, and the much larger 'toothed' Taemasosteus. Other arthrodires had jaws that could open wide, but they completely lacked any sign of teeth or denticles for catching large prey. The best example of this type from the Cavan region is Cavanosteus,[10] named after Cavan itself and the surrounding limestone. Cavanosteus is thought to have swum with an open mouth, filter-feeding on plankton, in the manner of the modern whale shark of today's oceans. Nearly fifty species of placoderms have so far been identified as having lived around the Devonian coral reefs that eventually formed the limestones extending westwards from Cavan and nearby Taemas all the way to Wee Jasper.

Here, it appears, swam a greater variety of vertebrates in the form of fish than anywhere else recorded in the Devonian world, including a group of strange, heavily armour-plated creatures known as placoderms.

—

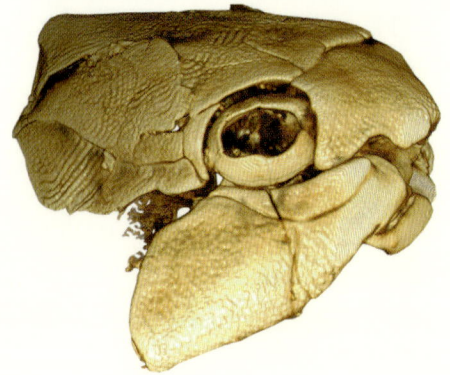

The exceptional number, array and quality of fossils that have been found in the limestone rocks are testament to the extraordinary importance of the Cavan area in geological and biological terms. Almost any piece of limestone will reveal numerous fossil remains of which at least sixty-two genera or groups of vertebrates have been documented as well as over 200 species of invertebrates. Moreover, some of the earliest known Devonian corals have been found in the limestone of Clear Hill and were first described in 1940 by Professor Dorothy Hill, one of Australia's most famous female scientists. Rarer still, however, are the fossil fish skulls that the region has produced (and continues to yield). In the late 1890s, a geology student from Sydney discovered a skull encased in limestone near the original Taemas Bridge, which proved to be the oldest fossil example from anywhere in the world of an ancient fish group known as the lungfish. These were more advanced than the placoderms; their chief characteristic was a lung that enabled them both to breathe and to remove waste. Five species of lungfish have been found to have inhabited the Taemas–Wee Jasper reefs. More recently, a new specimen, the largest lungfish jaw ever found, was removed from the limestone on another notable property to Cavan's west, Goodradigbee, which shares its name with the river.

Not only have the outer casings of fish skulls been discovered in the Cavan region, but there have also been several significant finds of uniquely preserved braincase structures of early

Whilst there are many Devonian placoderm specimens in museums and collections all over the globe, the only one that displays the entire skull, braincase, eye capsules, ossified jaw cartilages and tooth plates, was found on the hill just opposite the Cavan entrance.

—

vertebrates from 400 million years ago in a diversity of forms unequalled in any other fossil site on the planet. Importantly, no other fossil locality in the world has produced a perfectly preserved vertebrate eye capsule from a placoderm, one that includes nerves and muscle attachments controlling eye movement. Whilst there are many Devonian placoderm specimens in museums and collections all over the globe, the only one that displays the entire skull, braincase, eye capsules, ossified jaw cartilages and tooth plates, was found on the hill just opposite the Cavan entrance.[11] The value to scientists of this and the other finds mentioned cannot be overstated.

In 1939, some of these brain cases, fish skulls and other fossil specimens from the area were sent to the Natural History Museum in London, where they were used to pioneer what

Top left — *Bone of a Devonian placoderm; found in Cavan limestone*
Top right — *Perfectly preserved small placoderm skull showing the eye capsule, braincase and jaw; found on Narrangullen*
Centre left — *Perfectly preserved 400-million-year-old fish fossil eyeball of the placoderm Murrindalaspis; found on Cavan*
Centre right — *Jaw bone of the large placoderm Cavanosteus; found on Cavan*
Bottom left — *Reconstruction of a Devonian lungfish Speonesydrion, one of the oldest known lungfish species and found on Cavan*
Bottom right — *Skull bone from a large placoderm Taemasosteus; found on Taemas opposite Cavan*

— *Murrumbidgee River and Cavan Bluff*

has now become a standard technique of extracting fossil bone from limestone using acetic acid. This method was developed by a fossil preparator, Harry Toombs, and causes the surrounding rock to dissolve completely, revealing the fossils in three dimensions and in all their detail. The results of acid extraction were astonishing (especially because the specimens were so complete, the limestone having prevented the fossils from being crushed by the weight of sediment) and produced not only new species of fish, but new fish groups or genera previously unknown to science, as well as details of the interior of the vertebrate braincase that no one imagined could have been successfully preserved. Several hundred further fossils were collected in the 1950s for the same purpose and transported to London, where they form part of the research collections of the Natural History Museum and are recognised internationally.

CAVAN CAVES

Aside from its extraordinary limestone formations and fossils, Cavan was famous in the past for its caves, the locations of which never appear to have been properly recorded and have since been lost. According to the earliest account in 1832 by a contemporary, George Bennett,[12] the caves were 'very small and they did not repay the trouble of getting into them' in spite of there being many hanging stalactites. It seems, however, that Bennett may not have ventured far enough inside along what was apparently a rugged and difficult path, for he noted that 'the extent of the cavern was from 15–20 feet' with an interior so low that it was impossible to stand erect. Nonetheless, other accounts reveal that there were more chambers beyond the initial one and further wonders to behold.

A slightly later description on 8 November 1836 in *The Australian* is much more effusive. The anonymous author wrote:

The greatest attraction of the place [Cavan] are the caves ... The rock through which the cave is formed is black marble. The water which makes its way through the sides is strongly charged with some solution of lime, which forms the stalactites, the chief wonder of these regions. 'Ebden's Altar' is the most remarkable thing in this cave ... The whole of it has the appearance of alabaster. The altar is circular, and flat at the top, and about two feet in height; above it is what may be called the drapery, which has the appearance of aloe leaves hanging down within a few inches of the altar. By the bright blaze of a Bengal light ... the scene was romantic and unearthly, and the only sound which at first was heard was the rushing of the water through the cavern, astonishment having kept us all silent.

It seems Ebden's Altar had some spiritual significance for many of those who visited it, including William Riley, Cavan's owner at the time this article was published. In his will, he requested that he be buried at the foot of Ebden's Altar, a request that was ignored by his family who presumably did not share his feelings of reverence for the place.

The last known description of the caves is by journalist Thomas Walker, in 1838,[13] who praised them unequivocally. They were said by him to be seven or eight miles from the Cavan homestead and entered 'by a narrow fissure, a little way up the side of a hill, rising out of a deep valley and so covered by rocks that no one would be led to suppose there was any opening'. Walker described the network of caves to be found there in considerable detail, including some that were 'wide and lofty'

whilst others were 'narrow and low'. He too remarked on the beautiful stalactites, their 'fantastic shapes and appearances', and especially Ebden's Altar. He doubted that even half the caves had been explored, and it is frustrating that their whereabouts have not been handed down in folklore or oral history. Perhaps the fact that they apparently did not contain any fossils ultimately made them uninteresting to locals and visitors, who came for the greater prizes of organic remains to be found on the exterior of the limestone ridges.

THE INDIGENOUS COMMUNITY AROUND CAVAN

The ancient landscape of Cavan was not only at one time home to extraordinary fish, but millions of years later became the habitat of some of the first people to arrive in Australia. Cavan is in the very heart of Ngunawal country, an area that roughly encompasses Goulburn to the north, Gundagai to the west, Cooma to the south and Braidwood to the east. This also includes the entire boundary of the Australian Capital Territory and Canberra. Scientific evidence proves that the Ngunawal people lived here for more than 21,000 years prior to the arrival of Europeans. Comprising seven different clans that occupied fairly specific areas, the Ngunawal established one of the longest periods of continual habitation anywhere on earth.[14] This phenomenon was made possible by the Aborigines developing a way of living that was sympathetic to their environment and managing the natural resources of the land sustainably. The absence of colonisation and, in particular, of introduced animals and plant species, was, however, equally important to the longevity of Aboriginal occupation.

Around Cavan, the Ngunawal people would have had an abundant supply of wildlife and plants to feed themselves. They fished for perch in the waters of the Murrumbidgee using kangaroo grass to make fishing nets, and they hunted game with spears and boomerangs in the forests and on the more open plains or downs. These open spaces attracted kangaroos, wallabies, wallaroos, wombats, possums and emus, all of which formed the traditional diet of the Aborigine. Around the numerous swamps, lagoons and streams, there would have been plenty of aquatic birds to catch amongst the reeds, and lizards, snakes and native cats on the rocky limestone outcrops. Plant-based foods were also gathered, usually by women and children, and would have included yams, berries and grass seeds, the latter ground to make flour, which was mixed with water and then baked into a damper.

The Ngunawal were a self-sufficient and harmonious people with little need to travel far, though the different clans met from time to time around the area that is now Canberra for corroborees, trade purposes and marriage ceremonies. They also participated in the annual trek to the Snowy Mountains to feast on the Bogong moths, a great Aboriginal delicacy rich in protein. The gathering of various Aboriginal tribes would begin in October, after the snow had melted, when they would seek out the moths that had migrated to the mountain caves in order to escape the heat of their breeding grounds on the western slopes of New South Wales. Sticking to the rock walls like glue, the moths would be smoked out by the Aborigines before being cooked in ashes and eaten, their sweetish, nut-like flavour, drawn out by the heat.

Other than Aboriginal place names, there is little tangible evidence remaining of Ngunawal occupation in the Cavan region. The names of the three main watercourses in the area all derive from Aboriginal words – Murrumbidgee, meaning 'big water', Yass (originally *Yahr*), meaning 'running water' and Goodradigbee meaning 'water falling over rock'. Whilst these are obvious reminders of Aboriginal habitation, interaction between Europeans and the Ngunawal was recorded by Surveyor General Thomas Mitchell as early as 1836, when he passed through the Yass district in October of that year. He commented positively and with considerable warmth that,

My experience enables me to speak in the most favourable terms of the aborigines, whose degraded position in the midst of white population, affords no just criterion of their merits. The quickness of apprehension of those in the interior was very remarkable, for nothing in all the complicated adaptations we carried with us, either surprised or puzzled them. They are never awkward; on the contrary, in manners and general intelligence, they appear superior to any class of white rustics that I have seen. Their powers of mimicry seem extraordinary, and their shrewdness shines even through the medium of imperfect language, and renders them, in general, very agreeable companions.[16]

This peaceable experience of the indigenous population seems not to have been unique. Fortunately and somewhat surprisingly, there is no record of any massacres of Aborigines being carried out in the immediate Cavan area or wider district around Yass. There are few reported incidents of any conflict in the area at all and it seems that this particular Aboriginal community sought to establish itself in a viable position within the settler capitalist economy and society, by either working for the settlers themselves or, later, moving onto reserves that were created consequent to Aboriginal demands.[17] From the 1850s,

— *Native grasses in the Cavan district today*

as transportation and convict labour came to an end and the gold rushes created a temporary but severe labour shortage, Aborigines were rapidly recruited into the pastoral labour force, especially for shepherding and shearing.

With a relatively low stock density, some degree of compatibility was possible between Aboriginal and European land usage, even within the boundaries of a single property, and there are instances of groups of Aborigines living on a property whilst some members worked for the settler and others carried on their traditional subsistence activities. Sadly, there is no record at all of what the relationship was like between the local Ngunawal and the various inhabitants of Cavan, but no doubt as time went on and the property was improved with fencing, any Aborigines who were not directly employed at Cavan or on surrounding properties would have had difficulty surviving in the district. No accurate count of the Ngunawal tribe was ever made so it is impossible to say with any certainty to what extent white settlement was responsible for their ultimate decline; but undoubtedly, their numbers reduced due to the loss of their lands, and to the introduction of alcohol and diseases such as smallpox, syphilis and influenza caught from settlers.

They did not, however, become wholly extinct in the area; today, there are forty-two families around Yass that have a traceable Ngunawal bloodline.[18] This may, in part, be due to the fact that Yass became an experimental town for the New South Wales Aborigines Protection Board established in 1883. Initially, a number of reserves were established around Yass, including Oak Hill, Hollywood and Blacks Camp, reflecting government policies of protection and segregation prevailing at that time. Ultimately, these camps were unsuccessful; most Aborigines preferred to select where to live themselves, wanting to be close to schools and employment. A policy of assimilation

then ensued in the 1950s and many Aborigines from the reserves around Yass were compulsorily resettled in the township. Once again, this met with limited success; yet in spite of this and their complex and distressing history, the Ngunawal have survived in sufficient numbers to maintain a close connection with their traditional lands and be actively involved in the preservation of their culture.

ECOLOGY

The pristine native grasslands that prevailed in the Cavan district and around Yass at the time of Aboriginal occupation were critical to the flourishing of wildlife and Aborigine alike. This unspoilt country – subject nonetheless to deliberate periodic burning by the natives who understood such practices regenerated the soil and the herbage – was to come under huge pressure from 'the innumerable flocks and axes'[19] which the white settlers introduced. Agricultural methods in the British Isles, relied on by the colonists, could prove irrelevant or positively misleading under Australian conditions, and with few exceptions it took decades for graziers to learn how to manage their pastures in such a way as to 'harvest' instead of 'mine' the soil.

Before the arrival of white man, Cavan and its surrounds would have yielded in abundance grasses of varying heights, from the tall stalks of such species as kangaroo, wallaby, poa, spear and native sorghum, to a valuable lower storey of shorter grasses.[20] Many species, such as the shorter weeping grass, grew vigorously throughout the year, creating dense carpets of green. Other grasses remained dormant in winter, actively growing in hot dry summers. After the last frost, swards of kangaroo grass, covering the lower slopes in a rusty purple haze of dead leaves in

winter, turned green in the spring, their tall spikes rippling in the breeze across the hillsides, reminiscent of a cereal crop. It is no surprise that some early descriptions of the interior of New South Wales are unqualified in their praise; one contemporary wrote that he witnessed, 'Plains and open forest, untrodden by the foot of white man, and, as far as the eye can reach, covered with grass so luxuriant that it brushes the horseman in his saddle …'[21]

Amongst the grasses grew a multitude of colourful and perfumed flowers such as bulbine lilies of vivid yellow, deep purple fabaceae, everlasting daisies of white, pink and yellow, bluebells and a wealth of orchids. Tiny chocolate and vanilla lilies abounded, as did the Yass daisy, its roots providing nutritious food for the Aborigines. With the advent of spring, Cavan's abundant grasslands and woodlands exploded with colour and scent, the fragrant blooms lasting into early summer. Many varieties of shrubs also added to the splendour, from the startling yellow flowers of the wattle to the spidery, red blooms of the grevillea bush.

The grasslands of the Cavan district remained healthy and

regenerated by themselves because of their adaptation to their abiotic and biotic environments including their response to periodic burns instigated by the Aborigines, and because of the absence of heavy grazing. Trees, too, were plentiful, with a diversity of woodland and forest communities covering the mountains, slopes and valleys. Soil type, aspect, slope and altitude largely determined the species of trees in a particular spot; but kurrajongs, black cypress pines, native cherry trees and a wide variety of eucalypts, especially box, red stringybark, apple box and peppermint, grew throughout the region. Within many of these trees, flowers proliferated, especially in the eucalypts, drawing crowds of insects, birds and mammals to feast upon them.

Tragically, native pastures and woodlands such as these were degraded over time to such an extent by agricultural activities, climate change and government-regulated fire regimes that their spoilage now ranks amongst the greatest ecological disasters of the world. Sheep, in particular, have had a very negative impact on Australian grasslands which, prior to colonisation, had never been continuously and heavily grazed before. This landscape was not resilient enough to withstand the firm tread of sheep with cloven hooves, nor their grazing habit of eating grass down to the root – a very different treatment to the delicate footprint and light nibbling of the native marsupials. Nevertheless, with few exceptions, the ruinous effect of settlement in Australia on indigenous vegetation was not envisaged by contemporaries, and in the excitement of the discovery of fresh and verdant grazing lands, coupled with an increasing demand for wool, any warnings that were sounded went largely unheeded.

Above — *'Ribbons' of limestone clearly visible on Cavan*
Opposite — *Limestone hills along the Murrumbidgee at Cavan*

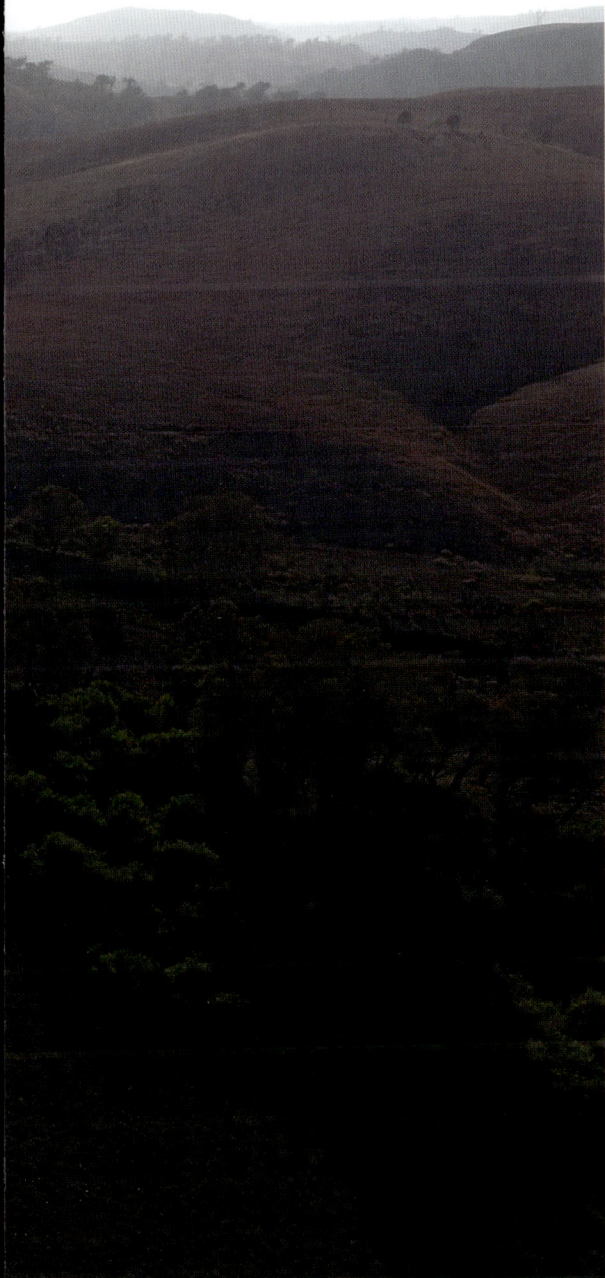

CHAPTER TWO

EARLY EXPLORATION AND SETTLERS

HUME AND HOVELL DISCOVER THE CAVAN DISTRICT

WHEN HAMILTON HUME AND WILLIAM HOVELL came across the Cavan district in October 1824 during their famous expedition from Appin (south of Sydney) to Port Phillip, what they saw were possibilities. These men were not concerned with balancing nature and human industry, however; they were explorers specifically commissioned by the Governor of New South Wales, Sir Thomas Brisbane, to identify new grazing land in the south of the colony. The natural pastures around Sydney and throughout the County of Cumberland (beyond which any occupation was illegal until 1825) had deteriorated to such an extent – due to indiscriminate grazing, overstocking and drought – that fresh pastures were urgently needed to provide existing and new settlers with a living. Until the Hume and Hovell expedition, the widely held view had been that the interior of New South Wales was an uninhabitable wilderness, suitable only for savages and wild animals.

Whilst Hume and Hovell are usually credited with the discovery of the Cavan region and especially the nearby Yass Plains, an immense tract of open country extending 25,000 to 30,000 acres, settlement in the area had in fact already begun.[1] A handful of enterprising settlers were dotted across the landscape the explorers traversed, occupying an area about 200 miles south and west of Sydney, and stretching as far as the Murrumbidgee on both its sides. Nonetheless, this was still frontier territory well outside the legal limits of occupation. Even when the

Nineteen Counties were declared in 1826, Cavan itself remained beyond any county boundary, lying to the west of the County of Murray across the Murrumbidgee. Not only does this set Cavan apart from many other early pastoral holdings, but as it lay in the path of the important Hume and Hovell expedition, its finding is well and truly anchored in Australian history.

Although the expedition was a great success, the record of it is fragmentary. Astonishingly, a final map of the journey was never published. Hume (possibly assisted by Hovell) traced their course on an expedition chart, provided by Sir Thomas Brisbane, but this has since been lost. One possible reason for the lack of a map was that the expedition was privately funded. Around 1830, Hume produced a sketch[2] reconstructing the route which, when taken with the written accounts by the two men, gives a reasonably accurate indication of the track the explorers followed. In fact, the Surveyor General, Sir Thomas Mitchell, found Hume's sketch 'surprisingly correct'[3] when he referred to it during his expedition up the Murray in 1836 to find the junction with the Murrumbidgee before returning to Sydney via the Yass Plains. It is this sketch and the written accounts of those on the expedition which determine conclusively that the northern tip of Cavan and the southeastern corner of neighbouring Narrangullen were crossed by Hume and Hovell on 23 and 24 October 1824.

On 19 October, a few days earlier, the expedition had reached the Murrumbidgee's northern banks directly opposite Cavan and found the river in flood. According to an account[4] by James Fitzpatrick, one of the expedition's assigned servants, although Hovell considered the river too high the party managed to cross 'by making Mr Hume's cart into a punt, taking it off the wheels, and covering the body of it with his tarpaulin'. Hume and another member of the expedition swam across the river with

Previous pages — *View over Cavan, looking east towards the new Taemas Bridge*

— *Hume's sketch map, dated 1830, of the route of his expedition with Hovell from Appin to Port Phillip*

— View over Murrumbidgee from Cavan

a fishing line clenched between their teeth, which they attached to the punt at the other end, to establish a connection between the banks. Eventually, 'with much trouble and not a little danger'[5] the whole group with cattle and stores were safely landed on the other side. They were enraptured with what they found. Hovell noted in his journal on 23 October, 'the whole is covered with a beautiful coat of excellent grass ... supperior [sic] to any I had seen in the Colony ... the whole has a fine appearance, and looks like Meddow [sic] land in England or that it had been Cultivated for Grasses.'[6]

A later account written by Hume in 1854, based on his notes of the expedition, records an incident on the 'Narrangullen

Meadows'[7] where the party camped on 23 and 24 October. Finding themselves hemmed in by the mountains, Hume went off in one direction and Hovell in another, seeking an outlet. According to Hume and another member of the party, Thomas Boyd, Hovell managed to get himself thoroughly lost for two days and was only reunited with the party after the firing of guns. Thereafter, according to Hume, Hovell 'never trusted himself out of sight of the camp, unless in my company'.[8] In subsequent, rather public arguments between them, Hovell refuted both this story and the fact that it was Hume's idea to convert his cart into a punt to cross the river. Even so, these accounts highlight the kinds of obstacles and difficulties encountered by the men during this gruelling expedition.

Most importantly, the mission to identify well-watered grazing land had been accomplished, and New South Wales was hailed as a promised land. As James Atkinson wrote in 1826, 'It is unnecessary for me to go into any detail to prove the excellence of the climate of New South Wales, its salubrity being well known and universally admitted.'[9] Atkinson, who had lived and farmed in the colony for several years, went on to praise, amongst other inland areas, the Yass Plains, describing them as 'generally a gentle undulating surface ... covered with grass.'[10] He maintained that the natural herbage and climate of the interior of New South Wales were so ideal for grazing and had aided the production of livestock to such an 'astonishing' extent that the colony 'justly claims precedence over many uncultivated countries'.[11] Without question, the meteoric rise in New South Wales in the 1820s of both sheep numbers and wool exported bore testament to this view.

They were enraptured with what they found. Hovell noted in his journal on 23 October, 'the whole is covered with a beautiful coat of excellent grass ... supperior [sic] to any I had seen in the Colony ... the whole has a fine appearance, and looks like Meddow [sic] land in England or that it had been Cultivated for Grasses.'

—

THE SPREAD OF SETTLEMENT TO CAVAN

Inappropriate pasture management and drought only partly explain the great exodus into the interior of the colony. Another significant factor was the increasing shortage of land supply around Sydney, which was a result of the rush of new settler arrivals in the 1820s and the impossibility for the vast majority of ex-convicts to return home. However, one of the main problems for those pioneers bold enough to venture further afield was a lack of security of tenure, as land beyond the County of Cumberland could not be granted, purchased or leased. Fortunately, in 1825, a ticket of occupation was introduced, which gave graziers occupying these outlying areas some certainty. A stockowner who had established himself in a suitable place could now obtain a ticket from the Colonial Secretary's office, giving him exclusive grazing rights to a particular tract of land, usually two miles in each direction from his stockyard, or for a specific number of acres not exceeding this amount. A ticket holder was permitted to cut only enough timber required to build himself a hut and a stockyard, and his right of occupation could be withdrawn by the Crown on six months' notice.

This system lasted little more than a year, when England ordered that the whole colony be divided into counties. As there were no maps available to do this, however, a limited area was chosen that could be surveyed and valued quickly; this area was designated in 1826 as the Nineteen Counties. Graziers and farmers were now expected to select land within the limits of these counties and to obtain title by grant or purchase. Adjoining land could be leased in addition, but settlement beyond any freehold or leasehold areas was discouraged. Those who ignored these boundaries were termed 'squatters' insofar as they occupied land without government permission. Their position was only officially regularised in 1836 by the *Squatting Act*, which introduced a system of licences enforceable by various commissioners for Crown lands. However, there is evidence that, as early as 1829, some grants were being issued for land that lay beyond the Nineteen Counties, though usually not very far away.[12]

JAMES FITZPATRICK

It seems that Cavan was one of these early exceptions, as it was situated outside the legal limits of occupation yet was granted title in 1831.[13] However, it is also possible that Cavan was occupied even before then. It has been suggested that James Fitzpatrick, one of the three assigned servants who accompanied Hume from Appin to Port Phillip in 1824, was Cavan's 'first settler'. Fitzpatrick was a big, strong man, a political prisoner

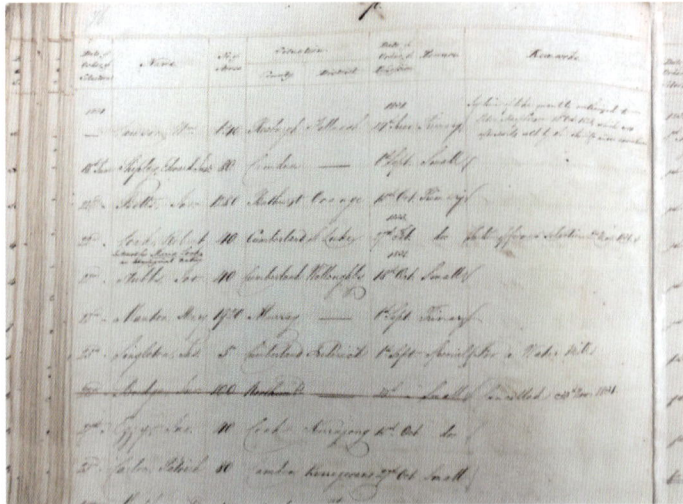

— *Entry in Colonial Secretary's papers showing grant of Cavan to Henry Manton*

transported for seven years in 1822 from Ireland for taking part in the Rockite Rebellion against the English and attacking a dwelling with firearms.[14] After his arrival in New South Wales on the *Mangles*, Fitzpatrick escaped but was recaptured and set to work on the roads before being given the chance of joining the Hume and Hovell expedition. Like many of the Irish prisoners at the time, Fitzpatrick was better educated and more intelligent than the average convict. In his diary, Hovell described him as 'a gentleman who in an unfortunate moment committed an offence for which he is enduring a punishment far too severe'.[15] With all that he may have been, however, he was not a good swimmer. A popular story still circulates that, during the expedition, when confronted with crossing the Murrumbidgee, Hume had advised Fitzpatrick to hang on to the tail of one of the bullocks. On arrival on the other bank, the bullock did what all cattle do when leaving water – it showered him with excrement, much to the amusement of the other men.

The claim that Fitzpatrick was 'the first settler' of Cavan came via a handwritten annotation on an old newspaper cutting[16] made by a much later occupier of Cavan, William Frederic Fletcher Castle-Roche. No other record survives to prove this assertion. Nevertheless, the few known facts about Fitzpatrick fit this conclusion. He was emancipated in 1825 as a reward for his contribution to the historic journey undertaken with Hume and Hovell, and it is possible that having been impressed by the lands he had seen, he returned, taking advantage of the ticket-of-occupation system and obtaining temporary but nonetheless exclusive grazing rights over what later became known as Cavan. These rights would have lapsed a short time later in 1826, when New South Wales was divided into the Nineteen Counties and the ticket of occupation was abolished. Perhaps he remained at Cavan as a squatter for a while longer before moving to the nearby Yass and Cootamundra districts, where he later held a series of depasturing licences (including one at Burrowa in 1838),[17] thus becoming a prominent landholder and member of the community.[18]

HENRY MANTON AND THE CAVAN GRANT

According to the remaining official record of occupation at Cavan, entered in 1831 in the Grants Register maintained by the Colonial Secretary, Henry Manton was the first settler promised a freehold grant of 1,920 acres (or three square miles). As was common practice at the time due to the pressure of applications, Manton first received a promise of the grant on or before 22 June 1831 and was then authorised to take possession of the land on 6 September the same year.[19] The actual grant, however,

— *Site of original Cavan homestead*

Presumably, Manton satisfied the authorities of his financial capabilities, though these could not have been very great given the setbacks his family had suffered in England. Henry was one of nine children of Joseph Manton and Mary Ann Aitkens, who married in 1794 at the fashionable St George's Church in Hanover Square, London. At the time of their marriage, Joseph was on his way to becoming the most highly respected and sought-after gun maker in England, widely acknowledged as a scientific and creative genius.[20] He was born in Grantham, Lincolnshire, into a long-established farming family that also ran a corn-milling business, but little more is known of his early background.

was not finally issued until eleven years later, on 27 July 1842, by which time Cavan had been sold.

The delay in the making of the formal grant was not unusual. The grants procedure was lengthy and cumbersome, requiring an applicant to write in prescribed form to the Surveyor General with a description of the land he had selected and details of his financial position. The Surveyor General then reported to the Governor to see if there were any objections to the selection and, if there were none, the Colonial Office issued a written authority allowing the applicant to take possession until a formal grant was issued. No settler received more land than the government considered he was capable of improving by means of his capital, and land was liable to be forfeited if it was not stocked and improved upon within eighteen months of a written authority to occupy.

— *Joseph Manton, c. 1810*

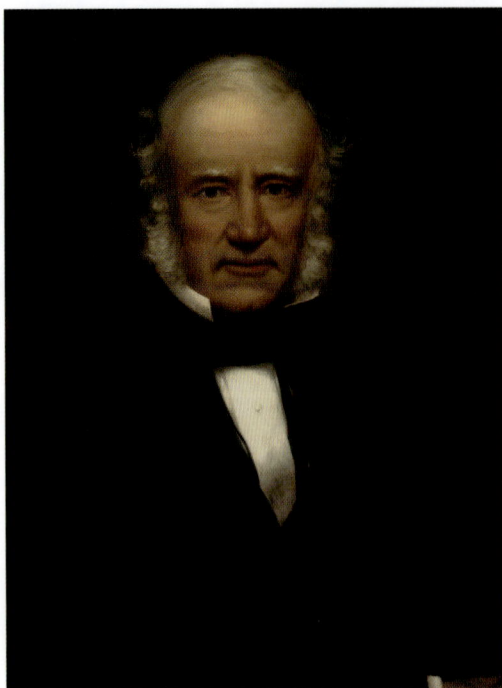

— Frederick Manton, c. 1850

Between 1780 and 1789, Joseph was apprenticed to a Grantham gunsmith and then to his eldest brother, John, who had a workshop at 6 Dover Street in Mayfair, London. When he struck out on his own, Joseph became not only more celebrated than his brother, but also world famous, climbing to the peak of his career in around 1810. Joseph was first and foremost a maker of fine sporting guns, whereas his brother specialised in pistols and general firearms. The secret to Joseph's success was that his guns would shoot faster than those of his rivals, therefore making his customers better shots. He achieved this by designing a special flintlock that fired much quicker than any developed by his contemporaries.[21] Joseph's guns were sought after by royalty, noblemen and gentry alike, commanding as

much as 70 guineas each. At the height of his fame, he was gun maker to George III and George IV, as well as to ten dukes and a large number of earls.

Unfortunately, flintlock guns were soon replaced by the percussion-cap lock, a design Joseph was not so adept at producing. He became embroiled in lengthy litigation for alleged violation of his own brother's patent and was undermined by his rivals who were able to produce percussion guns that were more reliable and cheaper than the superb flintlocks crafted under Joseph's exacting eye in his workshop in Davies Street near Berkeley Square. The business began to fail and Joseph was finally declared bankrupt in early 1826. When he died a pauper a few years later in 1835, his epitaph, written by his friend Colonel Peter Hawker, a distinguished soldier and one of the finest shots in England, declared him to be an 'unrivalled genius' and 'the greatest artist in firearms that ever the world produced … the father and founder of the modern gun trade'.[22]

It was into this interesting and entrepreneurial background that Henry Manton was born, on 23 March 1798. Whilst no information is available about his early life, entries in a cash book dated 1825[23] suggest that he joined his father in the gun business and was responsible for purchasing the silver and platinum required for the intricate decoration of the firearms. However, the business would already have been under threat when Henry was old enough to become involved. Henry was arrested for debt on 10 November 1825, along with his father and their clerk, and all three men were imprisoned at the King's Bench Prison in Southwark until their discharge three weeks later. It was shortly after this episode that Joseph was declared bankrupt. Although the Mantons struggled on, the business was finally closed in 1829.

— *Entry in Grants Register showing grant of Mon Reduit to Frederick Manton*

Prospects in England for Henry and his siblings were now bleak and it must have been about this time that Henry and his brother, Frederick, decided to emigrate to Australia. Frederick arrived in Sydney on 8 April 1829 on the *Guide* with his fiancée, Marie Émilie Blanchard. The couple later married and settled on a grant near Yass, which they called Mon Reduit. Henry arrived on 27 October 1830 on the *Lang*, though shipping records suggest he may already have been in Tasmania for some time before. Later, two other Manton brothers, Charles and John Augustus, also emigrated to Australia where they settled in Melbourne. History does not relate whether their father encouraged the boys to emigrate or if Henry and Frederick were persuaded by the numerous reports from New South Wales that farming of almost any kind was productive in so fine a country. Perhaps they engaged in pastoral pursuits influenced by their father's farming roots in Lincolnshire, but it seems from case reports in the local newspapers[24] that, like their father, three

out of the four brothers became insolvent. Henry gave evidence in the insolvency proceedings for his brother, Charles, and Frederick became the only one who prospered.

THE EMERGENCE OF A PASTORAL EXPORT ECONOMY

When Henry first settled at Cavan, the wool industry in New South Wales had already been revolutionised over the previous twenty years. Initially, when the colony had mostly been a gaol, sheep were kept not so that their wool could be exported, but to sustain the local convict population and its masters. There had been sheep in the colony since 1788 but, with the exception of John Macarthur,[25] no one in the early days had considered the possibility that great wealth could be gained from wool growing in a largely sweltering wilderness with only unskilled convicts for shepherds. Thus, sheep breeding was mostly indiscriminate and unscientific, with numbers multiplying at random. For the first settlers who had only small holdings of fifty acres to feed themselves and their families, their energy was devoted to the daunting task of survival.

Initially, when the colony had mostly been a gaol, sheep were kept not so that their wool could be exported, but to sustain the local convict population and its masters.

—

Once New South Wales began to produce wool that compared favourably with the best foreign wool and the British parliament reduced duty on colonial imports, Australian wool could triumph.

—

However, once the expansion into the hinterland of New South Wales was facilitated by grants and land sales in the late 1820s, creating large landholdings and landowners, the conversion of native grasses to wool on a significant scale became a reality.[26] Impetus was provided by an increasing demand for wool in Britain. This was due, in part, to a growing population, but also because of an interruption to wool supply from Europe during the Napoleonic Wars. British wool had become practically worthless in the 1820s because of competition from Europe's prime wool-producing countries, Spain and Germany, or more specifically, Saxony and Silesia. These countries were producing wool of exceptional fineness, softness, length and strength, that for reasons of comfort was favoured over the coarse wool produced at home and which, due to evolutionary changes in mechanisation, could be woven inexpensively into beautiful, lightweight cloth and other luxury materials. Once New South Wales began to produce wool that compared favourably with the best foreign wool and the British parliament reduced duty on colonial imports, Australian wool could triumph.

As one contemporary wrote, the whole of New South Wales was in a ferment of sheep growing: 'Men spoke of sheep and dreamed of sheep; they lived in an atmosphere of sheep unbelievable to one un-acquainted with the inexorable directness of colonial life and the unbounded optimism

of expectations in a new land.'[27] However, finding reliable labour and men with even a rudimentary knowledge of sheep husbandry was a significant problem, as was the lack of good stock to breed from. Nonetheless, by 1824, the wool industry had undergone a colossal transformation and the colony was exporting 400,000 pounds of wool per annum. By 1831, when Henry first arrived at Cavan, 2,250,000 pounds of wool were being exported each year, and by 1839 this had grown to an extraordinary 10,000,000 pounds per annum.[28]

With this explosion in demand and production, it is no wonder that Henry and Frederick decided to try their luck in the wool industry. Having no previous farming experience, it may be that they intended to farm together, especially as the Cavan grant was only fourteen miles from Mon Reduit, albeit on the other side of the Murrumbidgee. It is even conceivable that Henry obtained his grant on behalf of Frederick in order to overcome the strict limitations on the amount of land a grantee could be given (usually a maximum of 2,560 acres). Bearing in mind, too, that Cavan was sold only three years after Henry took possession, it's possible that in spite of the headiness surrounding wool growing in those days, he did not have any long-term ambitions to remain in the bush. No records survive as to Henry's life at Cavan nor what became of him after he left the district, except that he evidently moved to Sydney after 1834 and then to Melbourne before returning to England in the 1850s, where he died in 1857.[29]

Frederick, on the other hand, made a good life for himself on the Yass River in the wider Cavan district. His beautiful homestead was described poetically by the naturalist George Bennett in 1832 as 'being constructed on an elevated site' and commanding 'a fine picturesque view of the extensive plains or downs of Yas [sic], the distant wooded hills, forest scenery with

— *View from Cavan looking south across the early Cavan run*

the Yas [sic] river slowly winding its course beneath'.[30] Bennett continued, 'Mr Manton's farm is delightfully situated, having a fine stream of water running through it, every facility for sheep washing is afforded him – a desideratum of the first importance in this colony, where wool forms the staple article, the settler's main prop, and the cleaner it is brought to market, of course a better price can be obtained.'

Correctly predicting the importance of the Yass Plains for future pastoralism, Bennett concluded, 'This part of the colony appears valuable; the country is for the most part open forest, with luxuriant pasturage, and well-watered (an object of much importance in this arid country), combining capabilities of cultivation and grazing land with picturesque beauty. The "plains" or more properly speaking, extensive downs, are destitute of trees, affording abundance of pasturage for sheep etc, and the distance is terminated by open forest country, most part of which has already been granted or sold by government to settlers.'[31]

> 'This part of the colony appears valuable; the country is for the most part open forest, with luxuriant pasturage, and well-watered (an object of much importance in this arid country), combining capabilities of cultivation and grazing land with picturesque beauty.
>
> —

THE CAVAN RUN

In contrast to the Yass Plains, the land at Cavan was inferior, being rockier and less thickly covered with grasses; however, the area bordering the Murrumbidgee was exceptionally fertile and productive. The northern section of the Cavan grant lay directly along the river, but this was the narrower side of the property, the remainder of which stretched directly south along Mountain Creek. Whether Henry 'squatted' on or leased any adjoining land to his grant is not recorded, but the surrounding area came to be known as the Cavan run. This comprised the Crown lands that lay to the south and east of the Cavan grant, sweeping westwards towards Narrangullen as far as the Goodradigbee River.

Only estimates of runs were possible in this period as the land was unsurveyed, but it seems from later documentation that the Cavan run was approximately 51,000 acres.[32] With vague boundaries and little or no fencing, it became the site of many disputes amongst local settlers, vying for the use of the land. Until the Robertson Land Acts of the 1860s, which enabled the unimproved portions of the run to be broken up into separate blocks of conditional purchases, it was always an extensive area, distinct from the original Cavan grant.

As such, from 1828 it was legally available for grazing under lease to any adjoining landowner, and from 1836 to any persons on payment of an annual licence fee (subject to any existing lease), under the convoluted and ever-changing land regulations of New South Wales. Owners and occupiers of the Cavan grant almost exclusively depastured stock on the Cavan run, but on several occasions the run was 'lost' or allowed to lapse and was taken over by a third party. A prime example of this was a lease of the Cavan run to Fitzpatrick,[33] long after his supposed original appearance at Cavan and subsequent removal to Cootamundra.

From 1 January 1852, Fitzpatrick obtained a lease over the Cavan run for a term of eight years at £17 10/- per annum, although he was neither the owner nor occupier of the Cavan grant at that stage. The area was confirmed in the lease as comprising 51,000 acres of unsurveyed Crown land with a grazing capacity of 7,000 sheep. The lease was not witnessed nor entered in the official Register of Land Grants and Leases until March 1855, which may have been due to one of two reasons: either the completion of the lease was delayed by the usual inefficiencies of the bureaucracy at the time, when it was not uncommon for settlers' rights of occupation to be formalised retrospectively, or Fitzpatrick may have purchased the lease partway through its term from an existing lessee.

Although his name does not appear anywhere on the documentation, that lessee is likely to have been Horatio Beckham, the brother of Edgar Beckham, Commissioner for Crown Lands in the Lachlan district. Evidence for this is not only that Horatio Beckham had previously held depasturing licences over the Cavan run in the mid to late 1840s,[34] but it appears he also placed an advertisement in *The Sydney Morning Herald* on 7 January 1854 for the sale of the Cavan run and Marule Springs (another property in the Lachlan district about eighty miles away and known to have been leased by him at that time). No record of any sale pursuant to that advertisement has been found, but the fact that the lease to Fitzpatrick was formally completed after the advertisement and more than three years post its commencement date, is compelling evidence that a sale of the lease from Beckham to Fitzpatrick took place.

Fitzpatrick eventually quit the Cavan run in 1860 on the termination of the lease, and the land was once again taken

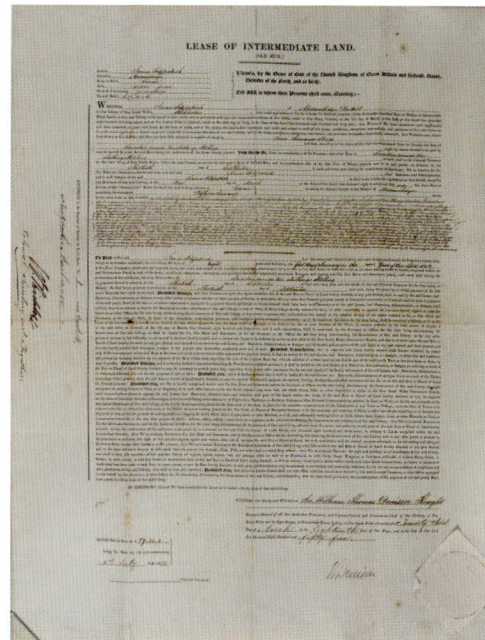

— *Lease of the Cavan run to James Fitzpatrick*

up by the then occupier of the Cavan grant. By this time, both the grant and the run had acquired a certain prominence as being part of the extensive landholdings belonging to the Riley family, whose name had become synonymous in Australia with excellence in sheep breeding – more particularly, the growing of fine merino wool. Their importation into New South Wales of large numbers of outstanding examples of Saxon sheep that became the base for a vast quantity of merino flocks in the colony was the catalyst for the Riley acquisition of both Cavan and nearby Narrangullen. It was on these properties that the Riley legacy evolved; with their highly scientific approach to sheep breeding, which challenged traditional, often primitive methods of wool growing, the Rileys were able to produce magnificent, world-class wool and change forever the nature of the colonial wool industry.

— *Dusk over Cavan*

CHAPTER THREE

ALEXANDER RILEY

THE CONNECTION TO CAVAN

ALEXANDER RILEY is widely known as one of the greatest pioneers of Australian pastoralism. His surname is often linked to Cavan, as it is one of the places where the Saxon merino bloodlines, for which he became famous, were developed. However, it was his son, William, who purchased Cavan from Henry Manton on 11 July 1834, one year after Alexander's death. Alexander never saw Cavan for himself, having returned permanently to London in 1817 following thirteen years' residence in New South Wales. Nonetheless, it is probable that father and son, who were in the wool-growing business together, discussed in correspondence the need for more land to graze their valuable Saxon sheep that by this time were transforming the Australian wool industry. These animals were already being kept at Raby, another Riley property nearer to Sydney, and also on Alexander's grant at Narrangullen, next door to Cavan, which made the purchase of the latter property an obvious choice. Whether or not Alexander knew about the prospect of purchasing Cavan before his death, it was undoubtedly his genius and drive that was behind the whole wool-growing enterprise at Cavan and across the other Riley properties.

FAMILY BACKGROUND

Alexander was a man of talent, courage, vision and enthusiasm. Born in London about 1778 (no record survives of his exact birth date), he was one of four children of George and Margaret Riley, who were married in 1777 at St George's in Hanover Square, the same church where Henry Manton's parents were wed seventeen years later. Like Henry Manton, the first grantee

Above — *Marriage certificate of George Riley*
Previous pages — *Alexander Riley's flagship property, Raby, by Joseph Lycett, 1826*

Alexander Riley is widely known as one of the greatest pioneers of Australian pastoralism. His surname is often linked to Cavan, as it is one of the places where the Saxon merino bloodlines, for which he became famous, were developed.

—

at Cavan, Alexander was the offspring of an educated and highly entrepreneurial father; it is perhaps no surprise that both Henry and Alexander emigrated, albeit at different times, to New South Wales, having had the example of initiative and risk-taking shown to them by their respective elders.

Many authors have stated that the Riley family were originally from County Cavan in Ireland, but no evidence has ever been produced to substantiate this and it must be assumed that the matter has always been one of conventional wisdom. Certainly, the surnames Riley, Reilly and O'Reilly are very common in County Cavan and usually attach to families that in the past were large landholders prior to English occupation in the seventeenth century. Alexander seems to have been proud of his roots, writing at one point to his nephew, Edward, that he hoped in the development of their business together Edward would 'consistently keep an eye extended to the fame of the O'Reillys!'[1] Alexander does not say whether or not these O'Reilly ancestors were from Cavan and, if so, when they changed their name to Riley, but there seems to be no other logical explanation for the subsequent naming of Henry Manton's grant which appears only to have been chosen upon Riley ownership. Even if the property had been named Cavan earlier than this, the source is unlikely to have been the 'presumed' first occupier, James

Fitzpatrick; although he was Irish, he came from County Limerick.

In spite of these facts, the origins of the Rileys and particularly Alexander's father, George, are very difficult to trace because relevant records were not mandatory at the time. From those records that do survive, there are three that might apply to him. The first is a George Reilly born in Cavan in 1744; the second is a George Rilley christened in York in the north of England on 8 April 1744;[2] the third is a George Riley (the spelling of whose surname is an exact match) who was born in London on 30 October 1743 and christened at St James's in Westminster, only a few miles from the area where George worked for all his adult life. His father is given as Daniel Riley, and perhaps it was he who had come from Cavan originally, though there is no evidence to support this.

Notwithstanding his obscure birth, George became not only a prominent bookseller, stationer and manufacturer of pencils and crayons,[3] but also, by a surprising coincidence with today's owner of Cavan, a newspaper proprietor! The two papers that George owned and published were the *Town and Country Herald* between 1799 and 1803, and the *Sunday Reformer* between 1799 and 1800.[4] Regrettably, no copies at all have survived of the former publication in any main library in the UK, and only twelve issues of the latter have been preserved at the British Library. Nevertheless, George's newspapers were amongst an estimated twenty-five dailies and weeklies that were circulating in London in the late eighteenth century, at least two of which had a readership of between 4,500 and 7,000.[5]

The *Sunday Reformer* seems to have been a weekly paper of wide interests, covering the arts, literature and entertainment as well as local and foreign news. This was a turbulent period in history, especially after the French Revolution, and newspapers played an important role in keeping the public informed about the British involvement on the various battlefields in Europe.[6] As its name implies, the *Sunday Reformer* also campaigned for social and political change, notably, the abolition of the slave trade in 1806. Technically, Sunday newspapers were illegal in this period, as only milk and mackerel could be sold on the Sabbath,[7] but the law was not rigorously enforced in spite of the outrage of many. Whilst George was never a newspaper magnate on the scale of Rupert Murdoch, perhaps he too was a man to draw controversy!

— Portrait of George Riley, c. 1799

No record has been found of George's early career or of any apprenticeship. The first immutable record is of him in London at Queen Street, Mayfair in 1770, when he would have been in his mid-twenties. He married Margaret Raby on 14 October 1777 and by 1781 had moved to 73 St Paul's Churchyard, which was in the most densely populated area of London's book trade activity, concentrated within the three-mile radius around St Paul's Cathedral. It has been estimated that in the late eighteenth century when George was in business, more than 4,000 people were active in the book trade in London alone, whether as publishers, stationers, ink makers, printers, bookbinders, engravers or other ancillary craftsmen.[8] According to one contemporary, James Lackington, London was the 'great emporium' of Britain for books, and bookshops stretched beyond the confines of the City of London to Fleet Street, the Strand, Piccadilly and Bond Street.[9] Frequently thronged with fashionable women, bookshops were considerable social centres, and many booksellers, including George, operated circulating libraries from their premises.

In common with many other booksellers in that period, George did not confine himself to bookselling activities. He was also famous for publishing children's books and educational games and selling historical playing cards and jigsaws.[10] By the late 1780s, he was selling pencils and thirty-two shades of coloured crayons, which he manufactured in Lambeth and advertised heavily in newspapers countrywide.[11] He even wrote and published in 1807 a *Concise Treatise on the Elementary Principles of Flower Painting* in which he explained the proper use of his special 'water cake colours' that he claimed would not fade and were better quality than rival paints, the colours being 'brought to such perfection' of 'chaste beauty' that they would 'please the eye of taste'.

Notwithstanding George's obvious entrepreneurial flair and enthusiasm for new undertakings, he was made bankrupt twice: the first occasion on 31 March 1778 and the second on 9 May 1801. Bookselling was very competitive in the late eighteenth century which is why so many booksellers combined other activities in an attempt to obviate hardships. It seems that the book trade in London alone increased by fifty percent in the last quarter of the 1800s and that there were 317 bankruptcies between 1772 and 1805 as a consequence of this increased competition. Whether George continued in business after his second bankruptcy is unclear, though he is recorded as being the proprietor of the *Town and Country Herald* until 1803, 'nearly the oldest proprietor of a newspaper in the kingdom'.[12] Nothing more has been found about him other than his death on 12 January 1829 in Greenwich at the age of eighty-six years.

EMIGRATION

Although back in England at the time of his father's death, George's son, Alexander, had earlier emigrated to New South Wales in 1804, becoming one of the first free settlers to the penal colony. He brought with him his new bride, Sarah Sophia Hardwicke, whom he had married the previous year. The impetus for this adventure may have been the earlier emigration of Alexander's two sisters, who had both married officers in the New South Wales Corps, Captains Ralph Wilson and Anthony Fenn Kemp. The latter was a scoundrel but nonetheless would go on to become the chief source of tobacco and spirits in Tasmania, president of the Bank of Van Diemen's Land, chairman of the movement to separate Tasmania from New South Wales, and one of Tasmania's most prominent citizens.[13]

— Frontispiece from George Riley's book, 1870

Perhaps inspired by Kemp, Alexander started out in trade, and soon after arriving in Australia he became a storekeeper and magistrate at the new settlement in Port Dalrymple, Tasmania, where his unscrupulous brother-in-law was second-in-command.

Finding favour with the authorities there, in 1805 Alexander was appointed Deputy Commissary, and in 1809 on returning

In a short time, Alexander became a key player in the early business enterprise of New South Wales and a merchant of note, forming a partnership with Richard Jones, another free settler from England. Together, they developed trading relations with Alexander's younger brother, Edward, who was living in Calcutta and trading in every possible commodity, shipping goods to Canton and the Australian colonies. Edward supplied Riley and Jones with regular cargoes of rum, clothing and food until he himself went to live in Sydney in 1816. However, Alexander appears not to have been content with merely importing goods for colonial sale and looked for a product to export to Britain that would bring him great wealth and position. It seems that it was almost by accident that he stumbled upon the staple he was searching for – merino wool.

— Alexander Riley as a young man (based on the portrait of his father on page 56)

BIRTH OF A WOOL-GROWING BUSINESS

to Sydney, Secretary of the colony of New South Wales. Both these positions gave Alexander insights into the vast profits that could be made from colonial trading and pastoral enterprise. Impatient for success, Alexander resigned his official post after only three months and began a long career of diversified commercial activities, dealing in an extraordinary variety of goods including clothing, tea, umbrellas and felt hats.[14] He even temporarily became interested in flax production, whaling and sealing; and was one of the joint contractors who, in exchange for a licence to import large quantities of spirits into the colony, erected the Sydney Hospital in Macquarie Street, popularly known as the 'Rum Hospital'.

In 1809, Alexander had received a grant of 3,000 acres near Liverpool, west of Sydney, which he named Raby after his mother's maiden name. With scant knowledge of farming, he stocked Raby with the best sheep available in the colony, a cross of the Cape and Bengal breeds with only a trace of Spanish merino. Shipping off the wool clip of 1811/12, Alexander was astonished to discover that some of it was worth 5/- 6d per pound on the inflated wool market in Britain during the Napoleonic Wars. These conflicts had created a disruption in the supply of wool from Spain and Saxony, causing a shortage in Britain where wool had been abandoned by many farmers in favour of corn or mutton production to feed the growing population.

Saxon wool especially was in short supply and sold for exorbitant prices, being recognised as the best in the world; it derived from some of the finest Spanish merinos which were exported to Saxony in the mid-eighteenth century and crossed with native sheep. However, the wool had been developed artificially by housing the sheep and rugging them up against the extremes of weather. This was thought by many at the time to be the only way of growing wool of such fineness, softness, strength and cleanliness that it would be unrivalled by any other fleece. Coarse British wools could not compete with this quality and, moreover, did not possess the strength in relation to fineness necessary for use with new mechanised techniques.

Unfortunately, however, the wool market collapsed at the end of the Napoleonic Wars and this, combined with a drought and restrictive land policies in the colony, temporarily deterred Alexander's pastoral endeavours. He returned to England with his family in 1817, leaving his brother, Edward, to replace him in his partnership with Richard Jones. On arriving in England, Alexander joined the experienced firm of colonial traders Donaldson, Wilkinson & Co, who handled large quantities of both New South Wales and German wool. Donaldson's son,

Stuart, was later to become embroiled in Riley affairs, with disastrous consequences for their financial wellbeing; but for the present, Alexander's new position gave him an extraordinary advantage.

THE IMPORTATION OF SAXON BLOODLINES

It was while working in the firm of Donaldson, Wilkinson & Co that Alexander realised he was uniquely placed for both growing and exporting fine merino wool to an unrivalled degree of efficiency and profitability. Not only did he have personal experience of wool production in Australia and all its vagaries, but he held a strategic position in the economic life of New South Wales through his brother, Edward, and the firm of Riley and Jones, as well as in the mercantile world of London. In short, Alexander was closely in touch with all the trends in both the British and New South Wales wool markets and possessed exceptional opportunities for arranging shipments from the colony and their sale and distribution in London.

Suddenly aware of the value of his landholdings in New South Wales, on which he was running 24,000 sheep, and of how growing and exporting wool in the colony could be profitable, Alexander evolved a strategy with his brother to bring the highly prized Saxon merino to New South Wales. It was this breed that ultimately became one of the two most important merino strains in Australian sheep breeding and transformed the whole of the colonial wool industry. Although the endeavour was a joint one, Alexander was its cautious trailblazer, amassing a huge amount of information regarding the husbandry of Saxon sheep and developments in the wool market in both hemispheres before

Saxon wool especially was in short supply and sold for exorbitant prices, being recognised as the best in the world; it derived from some of the finest Spanish merinos which were exported to Saxony in the mid-eighteenth century and crossed with native sheep.

—

— *Descendants of the Saxon merino at Cavan, the breed originally introduced into New South Wales by Alexander Riley*

making his investment. When Edward told him how land was being opened up to settlers to the south and west of Sydney, Alexander needed no more convincing.

Instructing an agent, William Dutton, to select sheep of high quality and productivity, Alexander purchased 198 Saxon sheep (20 rams and 178 ewes) for between £10 and £11 per head and loaded them onto the *Sir George Osborne*, chartered specially for the purpose. The cargo was shipped to Sydney in August 1825, with his nephew, Edward Riley the younger, in charge. Tragically, Alexander's brother, Edward, had taken his own life with a shotgun to the head a few months before, being prone to fits of depression and melancholia that were clearly not averted by the excitement of the new venture into Saxon sheep. Whilst the shipment was not the first flock of Saxons to arrive in the colony – that accolade belonging to Alexander's former partner, Richard Jones – it was nevertheless the largest, privately owned flock to be shipped there.[15] It was a huge gamble; Alexander described his outlay of capital as 'terrifying' but potentially enormously lucrative, as 'the sheep may be said to bear golden fleeces'.[16]

In charge of this precious cargo was the young Edward, not yet twenty years old, who, with a couple of shepherds, looked after the sheep with such care and dedication that, miraculously, only four died on the nineteen-week voyage. The diary of his adventure survives in the Mitchell Library in Sydney and records daily the health of the sheep and their diet. During the early part of the voyage, they were fed mostly hay and a feed mix made with water. Later, they were also fed oil cake, and if any were poorly they were given a bran mash. The only real problem Edward had with the sheep was heat stress, as the temperatures below deck hovered around 24 degrees Celsius or above. He resolved this in part by removing some partitions to allow for more ventilation. A violent storm on 24 August also caused some distress to the sheep, so that number 198 in particular 'seemed a little affected in his head'; but after a few weeks, Edward found them on 19 September 'all well and playing this morning like lambs in their pens'. Importantly, the flock included a poll ram (i.e. without horns) at Alexander's specific request – a move that pre-empted the development of the Australian merino by a century. The poll ram survived the journey, but later tragedy in the Riley family meant an end to the formation of what would have been Australia's first stud poll flock.

Shortly after landing the sheep at Woolloomooloo in Sydney, Edward took them to Raby where he stayed on to manage the flock for his uncle, Alexander. The latter had an anxious temperament and this trait, together with the need to protect his valuable investment, caused him to be a prolific and emotional correspondent. In many letters of often illegible and unrestrained handwriting, Alexander gave Edward meticulous instructions on how to care for the sheep and ordered him to keep the Saxons separate from any colony-bred animals. His

intention was to run a stud flock of Saxons, but also to ennoble his base flock by breeding these with the imported sheep, following a careful management programme. It is not surprising that Alexander was nervous about putting so valuable a flock in the hands of his inexperienced nephew; but the correspondence suggests that, whilst Alexander was a very effective business man, he was nonetheless a tortured one, worried about every detail and reluctant to delegate.

The serious depression in the wool market only nine months later served to fuel Alexander's anxiety; he wrote to Edward on 20 October 1826, alarmed about the collapse of several banks and trading corporations. He said, 'horrible revolutions have taken place in the mercantile affairs in this kingdom ... such

— *Letter from Alexander Riley to his son, William, 14 April 1830* times have not been known in England these last 100 years', and advised that 'wool has shared the general fate'. The tone of the letter is muted, but not entirely pessimistic. Whilst he was in London watching the wool price halve, he relied on the fact that the very best wool would still find a market. He cautioned Edward in the same letter not to overfeed the animals at Raby (a rather ironic instruction as drought conditions were prevailing at the time) and impressed upon him the importance of preparing, sorting and packing the wool according to the German method. He reminded Edward that the use of soft river water was the best for washing the fleeces, and advised him to lightly pack the wool in bales of 180 to 200 pounds to reduce the likelihood of deterioration on the long voyage to Europe and hence preserve its value on arrival.

Sitting out the wool crisis of the mid-1820s was doubtless agonising for Alexander and he worried constantly about money. Writing to Edward this time on 23 June 1827, Alexander was 'seriously alarmed at the large amounts' being debited against him for which Edward had apparently provided no clear explanation, and he told his nephew to spend at Raby only what was 'absolutely necessary for the preservation of the Saxon sheep'. Much to Alexander's horror, Edward had committed the cardinal sin of not only selling a number of imported Saxons to raise cash, but also crossing several of the Saxons with local breeds in a random and indiscriminate manner.[17] Luckily, things improved as the depression abated and, quite suddenly, Edward together with his cousin William began to do a booming business, selling the progeny of the Riley Saxons to colonial pastoralists. Indeed, they became so successful, relying heavily on advice from Alexander in London, that between 1827 and 1830, the Riley flocks won every gold medal awarded by the Australian Agricultural Society for sheep at its annual shows.[18]

THE INFLUENCE OF WILLIAM HAMPDEN DUTTON

It was not only Alexander who was behind the success of the Saxon sheep. He was fortunate to have the assistance of William Dutton, a brilliant agricultural scientist and widely acknowledged expert on Saxon merinos.[19] Dutton had been brought up in Saxony where his father had been British Vice Consul at Cuxhaven, an important harbour in the North Sea at the mouth of the Elbe. Also educated in Berne in Switzerland, Dutton spoke fluent German and at the end of his schooling in 1822 entered the only college in Europe at the time for agricultural science, the institute at Möglin near Berlin. Renowned agronomist Dr Albrecht Thaer had established the institute in 1804, at the command of King Friedrich Wilhelm III of Prussia who also gave Thaer the estate on which to build his college.

Thaer's institution became famous and, amongst other things, encouraged its pupils to apply science to the practical work of sheep husbandry, as taught by an experienced agriculturalist. There were also courses in chemistry, geology, mathematics, botany, entomology, veterinary science and agronomy so that an education at Möglin became a passport for managing vast estates and other agricultural ventures. It was so respected a qualification and so unusual in an Englishman that Dutton, who was naturally intelligent and talented, was considered by Sir John Jamison, President of the Agricultural Society of New South Wales, as 'the highest qualified authority' in England on the Saxon sheep and sorting of wool.[20] Not only was Dutton's training such that he was ultimately in a position to influence the genetic direction of sheep breeding in Australia, but he became one of the first men to emigrate to New South Wales with veterinary skills.[21]

Like Alexander Riley, Dutton was something of a visionary and entrepreneur. At the age of only twenty, he wrote to his father on 1 March 1824 that Great Britain could become independent of the Continent for her fine wool by 'erecting a national sheep farm in NSW and procuring a flock of the finest Electorals from Saxony' (a term denoting merino sheep from the Electorate of Saxony), using the Saxon rams to improve the local sheep. His remark was prescient; although at that time no sheep had yet been sent from Saxony to New South Wales, the matter was under discussion and the Australian Agricultural Company was formed in London a few months later, precisely with this aim in mind. This organisation was granted a million acres for the purpose and the venture was to be managed by an Englishman, Robert Dawson, a friend of the prominent colonial pastoralists, the Macarthurs. Dawson knew nothing whatsoever about conditions in New South Wales, although he had some experience of livestock management in England. He would need a sheep expert if the venture was to have any chance of success, and this role naturally fell to Dutton.

On 26 August 1825, Dutton was appointed by the Australian Agricultural Company as superintendent of their flocks for a term of seven years. He was sent to Germany and he purchased several hundred Saxon sheep from various sources, including from farmers near Leipzig and Dresden, though he also bought a number of sheep from Prince Lichnowsky of Silesia.[22] This nobleman owned the most famous flock of merino sheep in Europe that combined the superfine wool of the Electorals from Saxony and that of a larger framed type called Negretti, bred from the same Spanish merino bloodlines that had given Saxon sheep their almost unrivalled reputation.

Accompanying these sheep to Sydney, Dutton arrived with a total of 242 merinos on 22 March 1826, three months after

Dutton forecast that the Saxon sheep in Australia 'will give rise to a new character of wool, infinitely superior to the present' and that 'if proper attention be exercised in breeding the fine Saxons, we shall be able to transmit, from this country, a wool hitherto unknown to manufacturers'.

—

Alexander Riley had landed his first flock in the colony. Forty-two sheep were lost on the voyage and some were so weak on arrival that they died shortly after. Already previously landed in Sydney by the Australian Agricultural Company were over 700 Saxon merinos and more than 1,000 French merinos. In spite of losses amongst his own flock during the voyage, Dutton forecast that the Saxon sheep in Australia 'will give rise to a new character of wool, infinitely superior to the present' and that 'if proper attention be exercised in breeding the fine Saxons, we shall be able to transmit, from this country, a wool hitherto unknown to manufacturers'.[23] Ultimately, he was to be proved right, but not before a series of setbacks, the most important of which was disease.

Within a few months of landing, further sickness developed amongst the imported sheep that were being kept in a coastal area at Port Stephens, chosen by Dawson and his committee. The block contained large tracts of swampland and rugged mountain country, good training ground for horsemen but poor grazing for sheep. Here, they were very susceptible to diseases that flourished in a damp climate, whereas in their homeland they had been protected by being kept indoors on straw during

— *View of Cavan homestead and outbuildings looking northwest*

the winter months, which broke the life cycle of any worms. Dutton had wanted to move the sheep inland, but the company had not yet secured the necessary land. When sickness broke out, Dutton recorded the symptoms and the results of post-mortem examinations in such detail that veterinary scientists today have been able to conclude that the flock was heavily infested with pole worms in the lungs and intestines. These minute parasites were not known at the time and could therefore not have been diagnosed. However, the naysayers in England and Germany, who had been sceptical that Saxon sheep would be able to adapt to Australian conditions, were vindicated and Dutton was dismissed for negligence.

In spite of this ignominious parting, Dutton appears to have found permanent employment with the Rileys within two years. Possibly working briefly at Raby with Edward Riley, who was struggling to manage things on his own, Dutton returned to England where he subsequently came under Alexander Riley's full protection. It is not known how the two men originally met, but they were both clearly moving in the same mercantile circles in London. Alexander appears to have had great faith in Dutton, with his impeccable credentials, and had previously engaged him as agent in 1825 to select the first flock of Saxons that Alexander shipped to Australia. When these sheep began to prove a success, Alexander saw another opportunity; in spite of complaining of fatigue and exhaustion brought about by his volatile business affairs,[24] he turned to Dutton once again to select a second flock of Saxons for shipment to New South Wales.

In 1828, after Dutton had been sacked by the Australian Agricultural Company, he accompanied Alexander's son, William, to Europe with the express purpose of purchasing a further one hundred Saxon sheep, in part to repair the damage caused by Edward's sales and cross-breeding at Raby. The flock was of the highest quality, costing Alexander about £2000, and was shipped to Australia in August of that year. This time, Dutton accompanied the sheep on the voyage, in all likelihood together with William Riley, but returned to London shortly after, whereupon Alexander offered him a five-year agency agreement of which there is an extract in the Mitchell Library. This extract encapsulates the terms of a letter (now lost) dated 1 October 1829, authorising Dutton to go back to New South Wales and settle Alexander's affairs with his nephew, Edward, to value Alexander's lands in the colony, to supervise his interests there, to select 10,000 acres of land granted him by Lord Bathurst for his contribution to the pastoral industry, and to sell the Saxon sheep to the Australian Agricultural Company or manage them until otherwise sold.

There was by now a strong demand for Saxon rams in the colony that Alexander aimed to satisfy; moreover, Raby wool had become highly regarded in London. After Dutton had managed Raby for a couple of years with William Riley (who replaced his cousin Edward), the clip of 1832 that had been shorn, washed and sorted under Dutton's direction was of such outstanding quality that Alexander was able to write to William the following year: 'The fame of the Australian Saxon wools is now fairly established ... and I tremble in the possibility of any circumstance arising to prevent your sending this year's clip in the same state of perfection.'[25] Fortunately, Alexander's fears were never realised as the Saxons flourished at Raby, even though problems later arose between principal and agent that led to a premature parting of ways.

MANAGING THE SAXON FLOCKS

By the time Dutton arrived at Raby in early 1830, the Saxons were well established and had begun to bestow on Australia and the world at large 'a genetic legacy of unimagined proportions'.[26] They were also becoming extremely competitive. In spite of their long fleeces prone to greasiness and dust, it transpired that there was in fact no need for the exaggerated care the sheep received in Saxony. Beautiful wool could still be produced without supplementary feeding or the traditional rugging and housing in colder weather. By reducing the costs of production in this way, the Riley Saxon wool could compete readily with its German counterpart in English markets despite the difference in distance between the two production areas from Britain.[27]

These discoveries in the successful management of the Saxons can almost certainly be attributed to Dutton. He had been taught to apply science in the care and breeding of sheep during his studies at Möglin, and this is nowhere more clearly evident than in an extraordinarily detailed, unpublished three-part manuscript, entitled 'Sheep Husbandry', dated May 1828, and subtitled 'Development of Sheep Husbandry as practised in the best flocks of Saxony with particular reference to the growth of fine wool in New South Wales'.[28] Whilst anonymous, the first volume, when compared with a known sample of Dutton's handwriting, is highly likely to have been written by him, and its frequent reference to the work of Albrecht Thaer, the founder of Möglin where Dutton trained, points to the same conclusion. The date of the document is interesting as it was after Dutton's employment with the Australian Agricultural Company and before he accompanied the second shipment of the Riley Saxons to New South Wales. It seems that he wrote it (or at least began it) whilst in Europe with William Riley, selecting sheep for

In early 1830, the Saxons were well established and had begun to bestow on Australia and the world at large 'a genetic legacy of unimagined proportions'.

—

Alexander; this is supported by the partly illegible, but just decipherable, Prussian place name of Magdeburg that appears beside the date on the title page, indicating the place where it was written. William Riley is unlikely to have contributed to the treatise, as at that time he had not yet had any practical experience of farming in New South Wales.[29]

Importantly, the document explodes the theory which prevailed then that wool quality depended on climate, treatment and pasture. The author states that whilst there were different varieties of merino, even within Saxony, Thaer had shown that wool type was 'inherent in the blood' and that the desired qualities in a fleece could be perpetuated through the correct principles of breeding. In other words, what would otherwise be 'a casual frolic of nature' could be achieved consistently by selecting those animals with exactly the right characteristics, present to a high degree, and pairing them together. This was provided that a strict programme of discarding any sheep that deviated from the perfect model was followed. Significantly, these ideas of selection to create a specific type of merino came long before Darwin's theory of evolution.

The science of breeding is discussed in depth in the manuscript and shown to be far more important in achieving perfection in wool than any other factor. Above all, the author asserts that the selection of rams is the foremost consideration and that it is essential to know his origins to be sure of his

— A mob of Cavan sheep today

— *Showing Cavan wool's crimp and lustre*

powers of perpetuity. However, a warning accompanies this selection process: the greatest weight of fleece is never found combined with the best wool in the merino. The author continues, 'there is a point in which the quality may be so combined with the quantity of the wool in such proportions as to produce the greatest value of the fleece' and 'to hit this point and perpetuate it is the most difficult problem of the science of sheep breeding'.

There then follow chapters on gestation, birth, rearing, pastures, cross-breeding and registers, the latter of which enable the tracing of an animal's pedigree and are still in use today in the same format. Another 'modern' aspect of the treatise is its lengthy discussion in volume two of the wool itself and the build of the staple. Here, it explains that the fleece consists of an agglomeration of hairs known as the staple, which is found in numerous different forms. Whilst the fineness or diameter of the single hair is highly sought after, the author maintains that it is not by itself a guarantee of value as it can mean both brittleness and loss of weight in the fleece.

Of equal importance, according to the author, is pliability (softness and delicacy); elasticity or stretch; uniformity in diameter from the base to the top; and lustre, by which is meant a 'faint languid mild shine'. He claims that the most valuable wool is a dense fleece or 'the short stapled' variety which the sorters compare to a cauliflower 'because the washed fleece resembles this vegetable in compactness and roundness'. It is this ideal, 'this acme of perfection', that the author says all breeders should strive to attain to raise their wool above the common stock and ensure its value to the cloth manufacturers.

He states that such density in the wool (otherwise known as a close rather than open fleece) can only be achieved if the individual hairs are perfect according to the above characteristics. It also has the advantage of being less liable to 'external deteriorating influences' such as rain, dirt and dust, and of keeping the fleece whiter below the top. Produced only by sheep with finer hairs and thin skin, the latter being the result of 'the perfection to which such animals have been brought', the author explains that by parting a thick fleece with two fingers, the 'crowdedness of the hairs' and hence the probable weight of the fleece can be ascertained by the narrowness of the strip of red skin visible to the naked eye. In other words, the smaller the strip of skin, the denser the fleece, which in those sheep yielding the best quality wool, averages about three and a half pounds in weight.

The final chapters of the treatise cover washing, clipping and packing, the recommended methods for these procedures being adopted from those used in Germany. Apparently, the Rileys favoured a cold, live wash of sheep before shearing, allowing the fleece to soak fully, and then letting the animals stand for an hour or two in a nearby enclosure. The excess water had to be squeezed out of the fleeces by the shepherds by hand and, after shearing, the fleeces were rolled up in the German manner whereby six or seven were laid on top of one another, folded inwards and tied with string, before being placed loosely into a sack. The author writes, 'A slovenly manner practised in N S Wales of twisting the neck of the fleece into a rope and tying the wool up with it cannot be too highly censored'!

ALEXANDER'S OTHER INTERESTS

Implementing these innovative ideas on wool growing, especially careful breeding to achieve productive skin and fleece density, the Riley Saxons became pre-eminent in the early Australian wool industry. Nonetheless, Alexander was a restless man. Whilst breeding Saxon sheep was his principal focus, he was always on the lookout for other business opportunities, perhaps mindful of the frightening experience of the wool crisis of the mid-1820s.

About 1830, he began exporting cashmere angora goats to Sydney, having purchased these in France, apparently under the misapprehension that they were a type of sheep.[30] No doubt Alexander hoped they would prove an additional source of wealth, but they were more delicate than the Saxons, requiring yarding at night in sheds, and the small quantity of short, fine down which they produced was difficult to extract. In spite of being unprofitable, they were nonetheless widely admired and even exported to South Africa, but what eventually became of

them is not known. Another venture that Alexander tried was racehorses, which made him the first person to import an Irish thoroughbred, Skeleton, into the colony.[31] Although the stallion had been famous for his racing career, Alexander also gave this project up as a waste of time and money.

EXPANSION

Content then to pursue wool growing with single-minded determination, Alexander's Saxon sheep soon achieved such fame and desirability – because of the quality of the imported stock and the careful breeding and maintenance of standards – that Raby was not large enough to increase stock numbers and keep pace with demand. Almost 200 rams bred at Raby had been sold by 1831, and an early sheep classer, Ryrie Graham, stated that their imported sires 'were certainly the best I ever saw'. In fact, apart from some initial sales of rams by Richard Jones, surviving records show that, until the 1840s, only the Riley Saxons and those imported by the Australian Agricultural Company provided the colony with a continuing supply of Saxon blood. The Rileys had cornered the market and a search for more suitable land became essential to keep pace with the family's expanding business. With Alexander living in London, it was his son, William, who undertook this task with alacrity.

CHAPTER FOUR

WILLIAM RILEY

Above — *Cavan rams*
Previous pages — *Limestone ridges on Cavan, looking south*

SELECTING THE GRANT OF NARRANGULLEN

WILLIAM APPEARS TO HAVE ENJOYED a warm and affectionate relationship with his father. In a letter of instruction (the date of which is obscured)[1] prior to William's departure for New South Wales in 1828, Alexander adopts a very paternalistic tone. Amongst other things, he counsels his son to divide the day on board ship 'so that you may derive that benefit from your time which for months of confinement ... are so capable of affording you'. He advises keeping a daily journal and taking regular walks on deck and cautions William to 'avoid intimacies on board, especially with female companions'! Rather tenderly, Alexander writes, 'having once run my head rather hardly against the wall, I take you by the arm to prevent you doing the same'. With great respect for his son, Alexander recommends that on arrival, 'Take everything that is said to you by interested individuals with that discrimination which you are capable of exercising ... in forming your conclusions, exercise your own judgment and you will generally be correct.'

According to the surviving correspondence, Alexander initially intended for William and his cousin, Edward, to work together managing the family affairs in New South Wales with the support of William Dutton. It is not known how the responsibilities of Edward and William were divided, but Alexander remained nervous about Edward's inexperience and erratic management skills. In a letter dated 14 April 1830, Alexander wrote to William that after suffering for months in 'body and mind', 'I refer you, if not too late, to all my long communications to Mr Dutton which will fully possess you of the state of my affairs in NSW, and then you know sadly your father has been used, and unhappily at a time of life when least able to repair his injuries.' It is unclear to what these 'injuries' precisely refer and by whom they were inflicted, but the inference is that Edward's incompetence was at the root of these problems. The letter further highlights Alexander's perpetual anxiety about his pastoral pursuits.

However, the cousins seem to have been together at Raby for about eighteen months, presumably for a 'hand over' period before William began the search for more land to the southwest of Sydney. In 1825, Lord Bathurst had promised his father[2] a total of 10,000 acres in New South Wales made up of two separate grants of 5,000 acres each, conditional upon Alexander's success in introducing Saxon merinos to the country. For reasons that are not entirely clear, it took Alexander until May 1830, almost five years after the promise of the grant, to obtain formal permission from the authorities in Sydney to make his selection. It seems that neither Alexander nor his then agent, Richard Jones, did anything to push the application along, possibly because Alexander was preoccupied back in London with the collapse in the wool market, financial pressures generally and management issues at Raby that arose in this period. Once authority was granted, however, locating suitable land proved difficult.

According to a diary of his travels in August and September 1830,[3] which described in lively and poetic detail the countryside and its few inhabitants, William's first choice (and that of Dutton who accompanied him) was land on the Yass Plains. With his education in the care of sheep and the production of wool as learnt from Dutton, plus several months in Germany prior to his emigration, William was scathing of the flocks he saw on his journey. The sheep that belonged to an important landowner and grazier, Major Lockyer, apparently called for especially adverse remarks. William said, 'He has used

Riding around the open country near Yass, he climbed Cockatoo Hill and recorded that, 'Here we were greeted with one of nature's loveliest views; it was nearly sunset and around us we espied some 30,000 acres of the softest undulating plain or downs … beyond were the Murrumbidgee Mountains giving an air of grandeur to the whole … and I said to myself "this shall be my grant".'

—

some of Macarthur's old culls and it is not to be wondered at that his flocks do not improve.' Nevertheless, William was in raptures about the beauty of the landscape and its suitability for the Saxon sheep. Riding around the open country near Yass, he climbed Cockatoo Hill and recorded that, 'Here we were greeted with one of nature's loveliest views; it was nearly sunset and around us we espied some 30,000 acres of the softest undulating plain or downs … beyond were the Murrumbidgee Mountains giving an air of grandeur to the whole … and I said to myself "this shall be my grant and as far as Yahr [sic] Plains".'

However, such enthusiasm failed to take account of the complex land administration of New South Wales. As it transpired, the land William coveted was unavailable as it was within a church reserve. It seems that it was Dutton who tried to insist on this choice made on Alexander's behalf. In letters to the Surveyor General dated 30 July and 20 August 1831, he explained that he had been unable to find any other suitable land

and asked why the selection could not now be agreed, given that the lands in question had since been thrown open for authorised grantees. Dutton seems to have received no satisfactory answer, though there is a reference in later correspondence that the precise area he wanted had in the meantime been leased to an existing settler in the area, Henry O'Brien.[4] Clearly, Dutton succeeded in ruffling the feathers of the Surveyor General, who warned him that if he failed to make another unobjectionable selection by 1 October, the grants would be cancelled altogether!

Hastily, Dutton chose the land, which became Narrangullen, slipping in the selection just in time on 29 August 1831. The land was outside the Nineteen Counties and some thirty miles from where he and William had wanted to be on the Yass Plains (not eighty miles as Alexander contended when writing to Downing Street). Unwilling to let the matter rest there, Dutton and Alexander orchestrated a two-pronged attack on the authorities, with the former pursuing the matter in Sydney and the latter in London.

Writing on 6 February 1832 to the governor, Sir Richard Bourke, Dutton renewed his attempt to select land on the Yass Plains for his employer. He argued that in spite of having 'travelled upwards of 2,000 miles', he had been unable to discover 'a situation where I would venture to risk there a valuable property confided to my care'. He reminded the Governor that 'there is scarcely a stock owner in this territory who has not derived benefit from the introduction of the Saxon Sheep at Raby, indeed, this is a fact of such notoriety, that I can with confidence appeal to the colony at large for its correctness' and he spoke (presumably with tongue in cheek, given the views expressed in his treatise on wool [see Chapter Three, page 66] of the dangers of the climate to these 'delicate animals' as well as of the disadvantage of distance from Sydney). He finished,

'The natives are here exceedingly numerous and often so extremely troublesome to men and their stock, that I must not conceal from you the risk and danger of trusting in a situation so valuable a sheep as the Saxons ...'

—

with backhanded flattery, that 'a valuable flock like the Saxon Sheep cannot, in a colony like this, be a matter of indifference to its enlightened and liberal minded ruler'.

The Governor's answer has not been found amongst the correspondence but, suffice to say, Dutton did not succeed in his mission. Nor did his employer have any better luck in London. Alexander complained to the Colonial Office in a letter of 28 May 1832 that the land at 'Narrangullia' was not only some 80 miles beyond the spot originally applied for, but was also 'wholly ineligible and unsafe for the preservation and increase of such a valuable stock as the pure Saxons'. He quoted Dutton's warning (which in relation to the Aborigines differs from other contemporaneous accounts) that, 'The river over which it is only accessible is rapid and dangerous and sometimes for months cannot be crossed by sheep ... It is situated on the very limits of the western line of demarcation and the nearest magistrate is 80 miles distant.' No doubt, part of Alexander's objection was also due to the fact that this land was further away from Sydney, making transport more costly, but in the end, in spite of such misgivings, Narrangullen proved the perfect environment for the expanding Saxon flock.

Sadly, although William took possession of Narrangullen in 1831, his father was never to see his grants there (nor Cavan, which was adjoining and purchased later). Alexander died in London on 17 November 1833, having pursued to the last his utopian project of transforming the Australian wool industry into the finest in the world. By the time of his death, the Rileys had become the leading stud breeders in New South Wales and 'the foremost transposers in the colony of the Saxon tradition'.[5] The 200 or more rams that they had sold by this time exerted tremendous influence on New South Wales fine-wool flocks. Rams were also leased, an initiative of William's for which he charged £2 2/- per ewe,[6] and this only increased the effect on local stock. As William noted in his journal in 1830, some local flocks were so improved by breeding with his father's pure Saxon rams that 'the third cross might be mistaken for the Saxon wool having all the characteristics'. In short, over two decades of involvement by his family in the wool trade had left William in charge of a stud and wool operation that was a 'pacesetter' in Australia.[7]

THE FORTUNES OF WILLIAM HAMPDEN DUTTON

In the two years immediately preceding Alexander's death, it seems that Dutton, who had been so instrumental in the building up of the fine Saxon flocks on behalf of the Rileys, nonetheless proved to be a bitter disappointment. He was unreliable in rendering reports and accounts to Alexander in London and there is an implication in the surviving correspondence[8] that he was ultimately either negligent or dishonest, possibly diverting funds to finance his own pastoral

interests. None of these succeeded for any length of time, and he ended his days, financially ruined, as a wool sorter with a Melbourne wool-broking firm.[9] He died in 1849 at the young age of forty-four.

Dutton's opportunities, however, had been many. Aside from his potentially advantageous association with the Australian Agricultural Company, he had been employed by Alexander on somewhat unusual and magnanimous terms. It was agreed that he would manage Alexander's affairs in New South Wales in return for £100 for his 'passage and outfit', for an upfront loan for five years of £2,000 with interest at five percent (payable only in the last two years of his agency) and a life insurance policy. In addition, there may have been an annual salary or commission, but frustratingly only the first page of the employment terms survives. However, a copy of the promissory note for £2,000, dated 3 October 1829, is in the Mitchell Library, so presumably the five-year agency arrangement formally commenced on or around that date.

Clearly, at the time of his engagement, Dutton had ambitions of his own well beyond helping the Rileys. With his background of a gentleman and his expertise in sheep husbandry, he nurtured a dream to become a pastoralist himself and appears to have used the loan of £2,000 from Alexander to help him acquire his own landholding, Goodradigbee,[10] three miles west of Narrangullen. He took possession of this grant,[11] which comprised 2,650 acres, on 23 April 1831, having made the formal selection in October the previous year, presumably after seeing the land during his travels with William to the Yass Plains. As in the case of Cavan and Narrangullen, the grant at Goodradigbee was immediately outside the Nineteen Counties, occupying an area on the west of the Goodradigbee River at the junction with the Murrumbidgee – a very favourable position.

Cavan and Narrangullen were prime examples of how settlement was spreading in the 1830s and of the tacit agreement by government to allow occupation in unauthorised areas where limitless pastures and sun-baked plains lured the more ambitious and determined settlers.

—

But things did not go well for Dutton there. Tragically, the sheep with which he stocked Goodradigbee succumbed to a frightening disease which caused them to devour each other's lambs. This 'morbid appetite' of breeding ewes was fully described in gruesome detail by the naturalist George Bennett,[12] who explained that after eating sweet-tasting earth impregnated with alkaline salts especially prevalent in limestone country, the ewes would devour the progeny of the other ewes 'and, on the shepherds endeavouring to save the lambs just born from their voracity, they would rush upon them, biting their trowsers [sic], and making strenuous efforts to seize the lambs in the arms of the men'.[13] Bennett specifically mentioned that Dutton was one of the chief victims of this epidemic that also affected the Mantons at Cavan and Mon Reduit, as well as the O'Briens at Douro and the Humes at Humewood.[14] Dutton wrote to the government in November 1832[15] about his plight and asked if he could exchange his grant for land in another area, but this was refused in spite of his plea that the problem arose within four months of his occupation of Goodradigbee and rendered his selection 'useless'.

It was around the time that Dutton took possession of Goodradigbee that relations with the Rileys began to sour. The suggestion in correspondence is that Dutton had let Alexander down in some way, possibly defaulting on the interest payments on the loan of £2,000, especially after the collapse of his sheep enterprise, but also clearly failing in the management of Raby and Narrangullen. Unfortunately, the letters between Alexander and Dutton have not survived, but there is a letter from Alexander to William dated 16 March 1833 in which Alexander refers to Dutton having 'had at last transmitted accounts' and that 'I am sure you will long … have anticipated the disappointment and the grief they are so calculated to create'. Alexander goes on to say that Dutton incurred 'large expenditures' and 'I, at present, fear it must end in litigation with him'. Whether this is what happened has not been ascertained, but the agency arrangements came to an end shortly after in 1834, either as the result of breach of contract or upon the expiry of the agreed five-year term.

In the same year, after the calamitous episode of the 'morbid appetite' of breeding ewes, Dutton sold Goodradigbee as a cattle run, and moved to 625 acres near Yass, shown on old maps as Comur but which he renamed Hardwicke, presumably and rather curiously in honour of Alexander Riley's wife, Sarah, whose maiden name it was. A post office directory of 1835 and 1836 shows Dutton living at Hardwicke with his wife, Charlotte Cameron, the daughter of a lieutenant-colonel, whom he had married in 1831. Dutton seems to have been very much involved in the local community at this time, being appointed one of the first magistrates in the district in 1834 together with William Riley and Henry O'Brien.[16] Dutton took this role seriously, apparently hearing fourteen or fifteen cases each week, and campaigning for a gaol and new courthouse that was built in 1837.[17] He was outspoken about the number of squatters in the district, describing them as a serious evil, as many were recent emancipists and had no scruples about harbouring bushrangers, runaways and vagrants, or preying upon the stock and property of established settlers. As Dutton pointed out, 'The position was unavoidable – a kind of reflex of the convict-structure of society, especially in the newly settled outskirts where the law did not run.'[18]

But, unbelievably, further bad luck was to plague Dutton at Hardwicke. In 1835, his flock contracted catarrh, a condition similar to pneumonia, and he lost 4,511 sheep.[19] Dutton subsequently sold out in 1839, swiftly fastening onto the idea of becoming a pastoralist in the Mount Barker area of South Australia. According to a report in the *Sydney Gazette* dated 9 February 1839, he was 'so struck with the surpassing fertility of the Mount Barker district when compared to the richest tracts of New South Wales, that although [he] had arrived in the Province without the slightest intention of buying and or becoming [a] settler[s], had at once determined a special survey of 15,000 acres', not only to send herds and flocks there, but to people Mount Barker with newly arrived German migrants. This survey alone cost him the princely sum of £4,000, but by this stage Dutton must have been desperate after the miserable failures he had suffered in the Yass district. In spite of throwing himself wholeheartedly into the Mount Barker project, it seems that by 1841 he was up to his neck in debt and by 1846 was declared insolvent.[20] Like Alexander, Dutton was a utopian; but, in contrast to the Rileys, he lacked the Midas touch whether from ill luck, poor judgment or impetuosity, courting disaster almost everywhere he went.

— *Inside the new Cavan woolshed*

THE PURCHASE OF CAVAN

Without Dutton to advise him, William rose to the challenge of settling permanently in New South Wales and continuing his father's pastoral legacy. On 28 February 1833, he married the educated and intelligent heiress Honoria Brooks, one of six reputedly charming daughters of his neighbour at Raby, the prominent settler Richard Brooks of Denham Court. About this time, it seems that his cousin Edward faded out of the picture, his own father having died a wealthy man, leaving Edward sufficient funds to be independent and put his equivocal experience at Raby behind him.

William now proceeded to expand his landholdings, beginning with the purchase of Cavan from Henry Manton, the conveyance taking place on 11 July 1834 for the sum of £384. Cavan like Narrangullen was outside the recently formed Nineteen Counties and, instead, bordered the western side of the County of Murray on the opposite bank of the Murrumbidgee. It was another exception to the rule that settlers were supposed to occupy land within the 'limits of occupation' and highlights the enormous pressure the government was under to allow grazing in hitherto forbidden areas. Cavan and Narrangullen were prime examples of how settlement was spreading in the 1830s and of the tacit agreement by government to allow occupation in unauthorised areas where limitless pastures and sun-baked plains lured the more ambitious and determined settlers.

In 1835, William further extended his landholdings with the purchase of Warroo, 890 acres of valuable river country just north of Cavan, across the Murrumbidgee in the County of Murray. But it was at Cavan, not Warroo or Narrangullen, that the Rileys made their home and according to the earliest known

Above — *Entry in Grants Register showing William Riley's grant of Warroo*
Opposite — *View over Cavan homestead, looking northeast*

description, an article in *The Australian* dated 8 November 1836, the weatherboard homestead was a model of good order and housekeeping. The journalist admired it as 'the neatest and most gentlemanlike place' he had ever seen and went on to say, 'No logs of wood about, or broken drays, or ashes, but everything clean and handsome, with a flower-bed in front, and the soil raked and without weeds!'

FAMILY

Honoria bore William three children: Alexander Raby, born on 17 December 1833; Christiana, born on 17 July 1835; and Margaret Maria, born in April 1837. This youngest daughter was eventually to become the wife of one of Australia's best-known authors, Thomas Alexander Brown (known as Rolf Boldrewood), famous especially for his novel *Robbery under Arms*. Apparently, Margaret was no beauty, but she had bloodlines and a heritage that her husband found inspiring for his books. For Brown, who came from the younger colony of Victoria, she represented the world of old New South Wales, its glamour and its vicissitudes.

EXPANDING THE SAXON FLOCKS

Continuing his father's pioneering breeding of Saxon sheep, William kept Saxons at both Cavan and Raby. A sale notice that appeared in the *Sydney Gazette* dated 7 August 1834 is for Raby Saxon rams at Cavan. The notice states that 'A flock of about 300 pure Saxon Rams of the same character as those advertised for sale at Raby has been driven to Cavan on the Murrumbidgee where a boat will be kept for the purpose of conveying them across the river.' By this initiative, William ensured that his surplus stud flock could be easily distributed to Victoria and Tasmania, as well as within New South Wales and even as far as Western Australia. The following year, a similar notice dated 1 July 1835 appeared in the *Sydney Monitor* advertising for sale at Cavan another flock of 300 pure Saxon rams, 'equally remarkable' as those at Raby, with the same 'improvement in the size, length, fineness, closeness and weight of fleece'.

— *Margaret Maria Riley, youngest daughter of William Riley*

William was justly proud of the Riley flocks and, being a proficient sketcher, made a drawing of two of the prize-winning specimens at the Parramatta Fair as early as 1828.[21] Subsequently, this drawing was published in London as a lithograph, the fame and prestige of the Riley Saxons being well established there. According to the testimony of a well-respected sheep man, William Doughty, who classed the Riley Saxons at both Raby and Cavan, the sheep belonging to the famous pioneering family of the Macarthurs at Camden could not compete with the broad-backed and deep-bodied Riley flock,

— Poll merino stud rams at Cavan, descendants of the early Saxon sheep

— *William Riley's engraving of his prize-winning Saxon sheep, 1828*

THE WOOL INDUSTRY IN THE 1830s

This pre-eminent position was reached by following an exceptionally careful breeding, maintenance and wool preparation program, encapsulated in Dutton's treatise on sheep husbandry, and which was adhered to at Cavan as well as at the other Riley properties. Moreover, in the mid-1830s, wool from New South Wales was in huge demand abroad and the price was soaring. Australian wool sold on a buoyant market, not only to British buyers but, for the first time, to those in France and even Germany. Australia began to rival Germany as the principal source of supply for the growing British textile industry. The ignorance and experimentation in wool production of the early settlers in a strange, often intractable land were replaced with 'perfection' in both the breeding and management of sheep to such an extent that one commentator said, 'New South Wales may justly claim the admiration of every friend to industry and every lover of his country.'[23]

either in terms of their frame or density, or length and lustre of the wool.[22] Not everyone was satisfied, however, with the Riley Saxons, and one or two people complained about the quality of the rams, including Charles Cowper, Premier of New South Wales. But he was in the minority, and not only were the sheep sought after by other breeders, but the wool was highly regarded in London. Just before he died, Alexander wrote to William in July 1833, 'the fame of the Australian Saxon wools is now fairly established' and the Riley wool was especially regarded as 'most superior' by the whole trade.

SUICIDE

Apparently, however, these fortuitous conditions, an impressive inheritance, personal success and family were not enough to make William happy. Afflicted with the same gene that caused mental illness in his uncle Edward and acute anxiety in his father, he committed suicide at Cavan on 4 December 1836 at the age of only twenty-nine. The circumstances were mysterious and partly covered up by the circulation of conflicting stories about his death. One source published many years later claimed that he died from shock after witnessing the accidental firing of a canon at Cavan, which took out the eye of the prankster.[24] This seems unlikely as there were no contemporaneous reports in the local press of the alleged incident. A further story that gained currency was that William died after an incident at the first Yass races for which he had donated a cup, but these races were six weeks before his death and *The Australian* of 1 November 1836 recorded specifically that 'no accident happened to mar the festivities'. None of William's obituaries give the cause of death, except one that appeared in the *Sydney Gazette* asserting (incorrectly) that he died suddenly of apoplexy.[25]

More persuasive is William's 'will', or suicide note, scrawled immediately before his death and containing an emotional farewell message.[26] He wrote in a shaky hand, smudging the paper, 'On this evening of the third day of December 1836 the misery of my own feelings at my sad acts in moments of unexpected and extreme excitement have been so forcibly brought before me that I feel I cannot exist under such an impression of feeling.' The 'sad acts' to which he refers, nor the manner in which they were 'forcibly brought before' him that night, are not known. Perhaps he had been abusive or violent towards his wife or been unfaithful to her and could not live with the guilt. Curiously, he did not mention Honoria in his note at all even though she was pregnant with their third child. He bade farewell only to his friends, stating also that he wished to leave all his real property to his son, Alexander Raby, and all his personal property to his daughter, Christiana.

Since William made no provision for his pregnant wife and as the two children were minors, Honoria was obliged to challenge the will on the grounds that William was of unsound mind immediately before he died, 'labouring under a state of great mental excitement which was evident to other friends residing in the house'.[27] One of those witnesses was George Kemp, first cousin to William, who had been staying at Cavan for several

— *William Riley's suicide note*

weeks prior to the tragedy. If only he had left a record of what actually took place in those final hours and days!

William's request for a resting place at the foot of Ebden's Altar in the Cavan caves seems to have been ignored by his family and he was buried instead at St Peter's Anglican Church in Campbelltown, not far from Raby. Like his father who claimed that his 'eyes were ravished' on his first sight of Raby, William had loved his adopted country both for itself and the possibilities it presented for creating wealth. He was moved by the landscape in which he operated, rejoicing in so much natural beauty. Surveying the Goulburn Plains in 1830 during his search for more land to fulfil his father's grant, he wrote in his journal, 'Never was I more charmed, more astonished at the beauty and extent of the scenery which suddenly and unexpectedly burst on me ... and in the distance ranges of romantic mountains, it was quite magic to my eye.' Unlike many of his fellow settlers, William was not intimidated by the vast emptiness and oppressive silence of the Australian bush, but revelled in the plenty around him, adapting to colonial life with apparent ease.

THE RILEY LEGACY

Nonetheless, William's sudden and untimely death created an enormous and potentially disastrous vacuum. For a start, there was no one else in the family with the knowledge or skill to carry on the Saxon tradition at Raby, Cavan or Narrangullen; Dutton had left the Riley employment in disgrace, and William had also died in some debt. Reducing the number of Saxons seemed to be the best solution, and three auctions were advertised in the *Sydney Monitor* and *Sydney Herald* in April, May and July 1837.

> William's sudden and untimely death created an enormous and potentially disastrous vacuum.

—

A total of 170 Saxon rams and more than 350 Saxon ewes at Raby were sold, in addition to over 1,000 Saxon ewes at Cavan. Amongst the Cavan flock, there were 368 twelve-month-old ewes offered for sale, the first one hundred of which were bought by William's brother-in-law and friend, Edward Cox,[28] and taken to his stud, Rawdon. This flock had been founded on Raby blood and was run along the same breeding lines as all the Riley Saxons, with heavy culling and corrective mating and classing, so that the break-up of the Riley stud was not the calamity it could have been. At Rawdon and other nearby Mudgee studs, the Riley Saxons continued their influence and extraordinary contribution to the wool industry.

Despite these sales, the bulk of the Riley flock initially remained at Raby and Cavan. However, when Cavan was leased a short time later, it seems that a number of Riley Saxons were sold to the new tenant, Major Edmund Lockyer.[29] The remainder of the Cavan flock (and also that from Raby which had also been leased) was then moved north by the Riley executor, Stuart Donaldson, to his own properties at Tenterfield and Clifton in northern New South Wales, where they were sold in the mid-1850s[30] and subsequently vanished into obscurity. Thus, whilst the original Riley Saxon flock was finally dismantled altogether, its bloodlines lived on, as did the Rileys' innovative approach to breeding and wool preparation, passed on to some extent by friends, experts and ordinary men who had worked on the Riley properties.

Cavan still drew praise after William's death, although the accommodation was surprisingly modest. A very comprehensive description of Cavan can be found in Thomas Walker's *A month in the bush of Australia*, published in 1838:[31]

When we had dined, the gentlemen walked out to examine the improvements etc about Cavan; these are not very extensive. The house itself is a small weather-boarded cottage, with neat well proportioned and well finished rooms, and does not look like the common order of bush cottages: behind it are the offices, in a line perpendicular to the length line of the house: at a little distance are the stables, and at some further distance, on the flat, is a good, but not large, brick-built barn just finished ... The cultivated land is very rich limestone and alluvial deposits, bounded on one side by the river, and comprises 150 acres I should say. There are other paddocks fenced in. There is a garden near the cottage. The place altogether has an appearance of neatness and taste about it ... The view and country around are very picturesque and pretty, though not seen to most advantage from the cottage ... They consist of the banks of the Murrumbidgee River, here a running stream, with a good deal of water in it, even at present, flowing circuitously through a very broken country. There are a few flats, but the country is chiefly very hilly on each side, and bounded by rather high ranges; the hills are, however, very thinly timbered, and covered with grass, so that the whole view ... of river, meandering over a stony bed, through here and there flats, with hills and ranges such as I have described, is very beautiful indeed; notwithstanding all that, it was not such a place as I should like to live at, there is a wilderness about it, or something else, that prevented me from feeling that it was a place I should choose for my

residence; it is rather too extensive to be considered a snug sheltered valley, and yet it is nothing more than a valley of broken though beautiful ranges.

From this description, it seems that Cavan was perfectly well managed after the death of its proprietor, but had an air of desolation and melancholia about it, doubtless caused by the tragedy of William's suicide and Honoria's subsequent move to Sydney. There, she had set up a fairly elaborate establishment for herself, which housed a large collection of books, and drove around town in a London-made phaeton drawn by some 'excellent' horses.[32] She herself does not appear to have had either the inclination or business skills to fill her husband's shoes, although she could presumably have played some kind of management role at Cavan given that her son, the heir, was still a minor. Instead, however, Cavan was allowed to drift during a protracted family dispute over the validity of William's will and the appropriate administration of his affairs. The future of Cavan thus hung in the balance, as did the fate of the Saxon sheep.

53.	Reg. No.	Fol.	Grantee	Quantity	Tenure	County and Parish	By whom Granted	When Granted	Annual Quit Rent	When it comm. renewed	Name of Farm	Witness	
187	76	61	James Richard Styles of Parramatta.	640	Land purchase	Murray, Unnamed, near Bundarra.	Sir Geo. Gipps	4th May 1842.	One farthing forever			Sir Geo. Gipps.	Commencing at ... Dyce's purchase ... North by that line ... Styles one thousand ... eighty chains; ... chains, and on the ... chains to the Gum ... purchase. Being ... of 23rd March 1...
188	71	261	Alexander Raby Riley of Raby.	1920	Grant of Land, B.	Unnamed, Unnamed, at Cavan.	Sir Geo. Gipps.	27th July 1842.	£16.0.0	1st January 1839.		Sir Geo. Gipps.	Commencing at a ... the East by a line ... a line bearing West ... and thirty seven ch... river to the marked ... land promised to Herr... Sir Ralph Darling ... 1834, as a primary Grant... ance with the report ... appointed under the ...
						Note — This Grant is situated nearest to the County of Murray.							
189	76	101	Patrick Hyill of Parramatta, in trust.	4000	Land purchase	Murray, Unnamed, at Bungendore near Lake George.	Sir Geo. Gipps.	4th Novr 1842.	One farthing forever			Sir Geo. Gipps.	Bounded on the W... at the South East cor... East two hundred an... creek to the Village re... seven chains West eigh... and on the North Ea... additional Grant af... made on the 15th day of ... the Colonial Legislat... upon the trusts of the ... the 26th day of Augu...
190	71	275	Alexander Raby Riley of Raby.	5000	Grant of Land, B.	Unnamed, Unnamed, at Murrumgullen.	Sir Geo. Gipps.	14th Septr 1842.	£41.13.4	1st January 1838.		Sir Geo. Gipps.	commencing at the ... five thousand acres re... ing South five hundred ... sixty four chains; N... chains to a Creek, and ... and East by that li... sand acres Reserve ... Riley, on or before the ... ance of instructions p... authorized to take poss... to the said Alexander ... on the 28th June 18... nual Legislature 5...
						Note — This Grant is situated nearest to the County of Murray.							

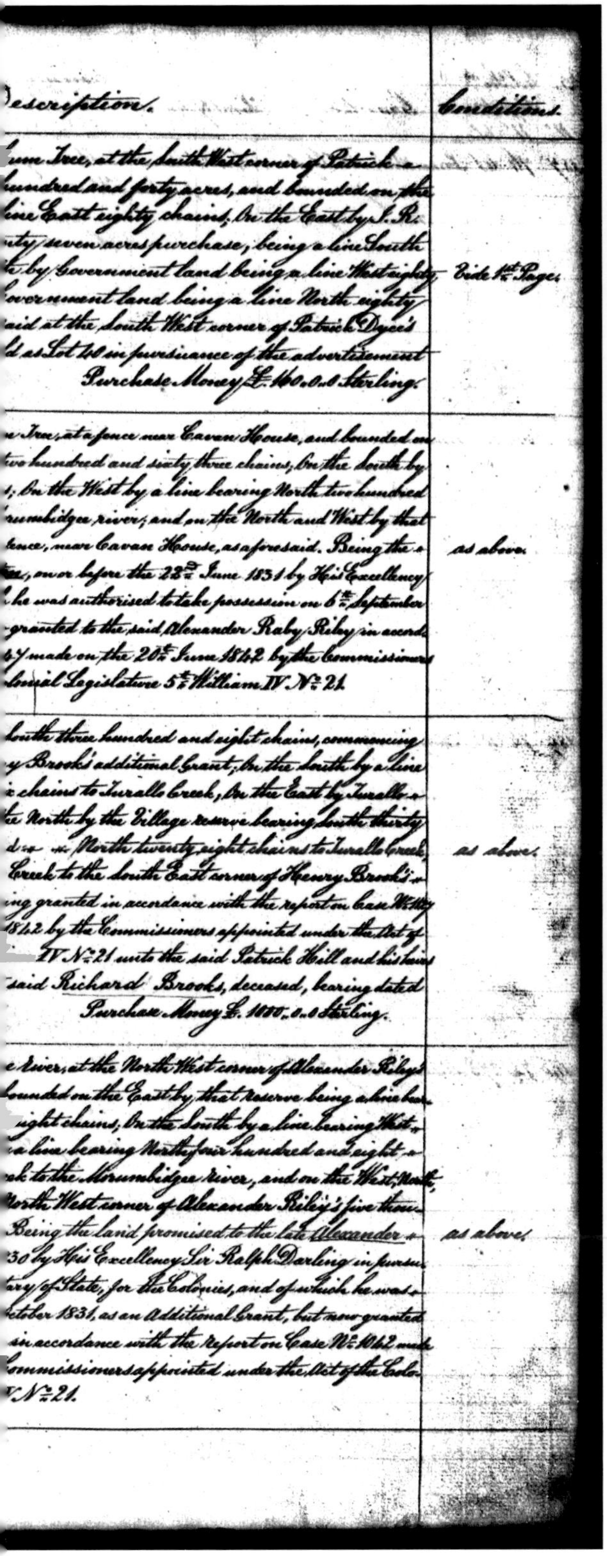
escription. | Conditions.

um Tree, at the South West corner of Patrick a
hundred and forty acres, and bounded on the
line East eighty chains. On the East by J. R
nty seven acres purchase, being a line South
h by Government land being a line West eighty | Vide 1st Page
overnment land being a line North eighty
said at the South West corner of Patrick Dyce's
ld as Lot 40 in pursuance of the advertisement
Purchase Money £. 140.0.0 Sterling.

n Tree, at a fence near Cavan House, and bounded on
two hundred and sixty three chains, On the South by
; On the West by a line bearing North two hundred
rrumbidgee river; and on the North and West by that
ence, near Cavan House, as aforesaid. Being the | as above.
, on or before the 28th June 1831 by His Excellency
he was authorised to take possession on 6th September
granted to the said Alexander Raby Riley in accord
made on the 20th June 1842 by the Commissioners
lonial Legislature 5th William IV No. 21.

South three hundred and eight chains, commencing
y Brooks' additional Grant, On the South by a line
chains to Jeralla Creek, On the East by Jeralla
e North by the Village reserve bearing South thirty
North twenty eight chains to Jeralla Creek, | as above.
Creek to the South East corner of Henry Brooks'
ing granted in accordance with the report on Case No.
1842 by the Commissioners appointed under the Act of
IV No. 21 unto the said Patrick Hill and his heirs
said Richard Brooks, deceased, bearing date
Purchase Money £. 1000.0.0 Sterling.

e river, at the North West corner of Alexander Riley's
bounded on the East by that reserve being a line bea
ight chains, On the South by a line bearing West
a line bearing North four hundred and eight
ek to the Morumbidgee river, and on the West, North
North West corner of Alexander Riley's five thou
Being the land promised to the late Alexander | as above.
30 by His Excellency Sir Ralph Darling in pursu
ary of State, for the Colonies, and of which he was
ctober 1831, as an Additional Grant, but now granted
in accordance with the report on Case No. 1042 made
ommissioners appointed under the Act of the Colo
V No. 21.

CHAPTER FIVE

INTERREGNUM

THE DEATH OF HONORIA RILEY

LESS THAN THREE YEARS after William Riley's death, his widow, Honoria, died on 17 March 1839, having never recovered from the sudden loss of her husband. Perhaps she was also disheartened by the lack of provision for her in his will, and by the dispute that subsequently arose between herself and her brother, Henry Brooks, over who should administer the Riley estate in the absence of any executors. Unlike her late husband, Honoria left a formal will[1] which stipulated that she wished her children to be educated, kept together and cared for by her sister, Christiana Blomfield, who was living with her husband and family at the Brooks's ancestral home, Denham Court. She also requested that her sister receive the sum of £300 per annum, 'for the maintenance and comfort of my three children until they are respectively of age or marry'. On the face of it, this dying wish should have been relatively easy for the Riley estate to satisfy, given not only the extensive landholdings that included Cavan, Narrangullen and Raby, but also the money raised from the sale of Saxons shortly after William's death and the income produced by the valuable flock that still remained. However, the Riley affairs were neither transparent nor straightforward after Honoria's death and did not provide the stability or inheritance that Honoria had clearly anticipated for her offspring.

Above — *Portrait of Stuart Donaldson by G.B. Shaw, 1856*
Previous pages — *Entries in Grants Register showing grants of Cavan and Narrangullen to Alexander Raby Riley*

TENANTS AT CAVAN

Before she died, Honoria was unwell for many months; although the courts finally granted her (and not her brother) the administration of her husband's estate, it seems she was too weak to carry this out, so she gave a family friend, Stuart Donaldson, authority to manage her affairs for her. Honoria died not long after, and the courts formally appointed Donaldson administrator of William Riley's estate.[2] A little younger than the late William Riley, Donaldson was the son of one of the founders of Donaldson, Wilkinson & Co, the mercantile firm in London with which Alexander Riley had been associated for many years. With this connection, it is highly likely that on being sent to Sydney in 1835 to stimulate business, the young Donaldson arrived in the colony with a letter of introduction to William and Honoria. Having made their acquaintance, it seems Honoria took Donaldson into her confidence, appointing him joint executor of her will with her brother-in-law, Thomas Valentine Blomfield, and also guardian of her children. Prior to this, however, Donaldson took the decision, apparently in his capacity as an informal adviser to Honoria, to lease Cavan and Narrangullen in 1838 to a retired army major, Edmund Lockyer, on behalf of his son-in-law, Captain George Potter.[3]

MAJOR EDMUND LOCKYER

Lockyer was a very suitable choice of tenant, since he was well respected in colonial circles and also a person of substance. He was born on 21 January 1784 in Plymouth in England, the son of a sailmaker, and entered the army at the age of nineteen. After serving in England, Ireland, India and Ceylon, he was sent to Sydney where he arrived in April 1825 with his wife and ten children. By this time, he was a major, and after several important assignments on the east and west coasts of Australia, he sold his commission in 1827 and obtained a grant of 2,560 acres in the Marulan district, northeast of Cavan, which he named Lockyersleigh. By 1837, he had added

— Major Edmund Lockyer, c. 1840

another 3,635 acres to Lockyersleigh by purchase, and by 1853 the estate totalled 11,810 acres.[4] In short, by the time Lockyer leased Cavan and Narrangullen from the Riley estate, he was a significant landholder and may even have been personally acquainted with the Riley family through business or social connections. This did not, however, guarantee the quality of his livestock. When William Riley passed Lockyersleigh on his journey to the Yass Plains in 1830, he commented disparagingly in his journal that 'the Major is happy in possessing 5 flocks of miserable, coarse sheep'![5]

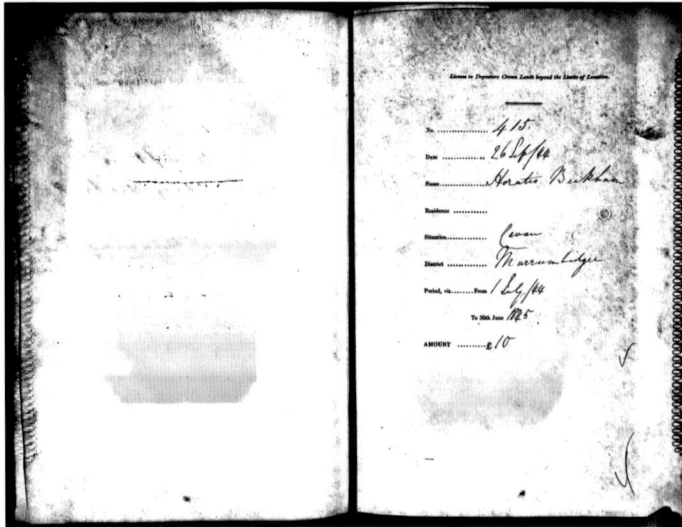

— *Depasturing licence over the Cavan run in favour of Horatio Beckham*

Upon taking the lease of Cavan, it seems that Lockyer also obtained a depasturing licence over the Cavan run, a form of permission granted by the Crown to graze land both within and beyond the Nineteen Counties in return for an annual fee of £10. Whilst these licences recognised occupation by pastoralists, the tenure was by no means secure because the licence was for one year only, improvements were made at the occupier's own risk, the land was unsurveyed and, most importantly, the boundaries were unclear and subject to negotiation with neighbours. According to a letter dated 17 April 1844 from Stuart Donaldson to the Colonial Secretary,[6] the Cavan run of about 51,000 acres had formerly been used by the Rileys for grazing sheep; but during Lockyer's tenancy, he (or Donaldson as executor of the Riley estate) neglected to renew a licence expiring on 30 June 1842. This resulted in a substantial part of the run being lost and subsequently made over to Horatio

Beckham, a brother of Edgar Beckham, Commissioner of Crown Lands for the Lachlan district.[7]

A huge fuss ensued[8] on the grounds of nepotism, since the official responsible for granting the licence was Henry Bingham, another Commissioner of Crown Lands, but for the Murrumbidgee district. The argument was that Edgar Beckham had used his position to secure the licence for Horatio, that it had not been made to someone 'at arms' length' and that Horatio was probably acting on behalf of Edgar. Whilst commissioners were not prohibited from obtaining these squatting licences for themselves, the practice was nevertheless frowned upon. An enquiry was held in which Bingham argued[9] that he had acted properly by initially reserving the land for Lockyer who he knew had met with 'some severe losses' at the time (probably due to the prevailing downturn in the pastoral economy) and who had contacted him shortly before the licence expired to advise him that sheep were on their way from Lockyersleigh to the Cavan run. However, on hearing nothing further from Lockyer in spite of several letters drawing his attention to the vacant land, Bingham felt he was justified in granting the licence application to Horatio Beckham a year later on 1 July 1843, considering that he was 'bound to see that his Majesty's Crown Lands were duly occupied and that under the circumstances, neither Major Lockyer nor any other trustee or agent for the estate could have, after such a length of time, any claim to the lands'.[10] Moreover, Bingham pointed out that Lockyer's son-in-law, George Potter, 'resided constantly at Cavan and had frequent opportunities of seeing me personally if he wished to do so' to arrange renewal of the licence in question.

The government agreed with Bingham and restitution of the lost portion of the run was refused. There were then several years of disputes during which it was alleged that Beckham's sheep were allowed to graze so close to Potter's cattle that the latter were being scattered all over the country, whereas Beckham complained that Potter ordered his shepherds to allow his sheep to go where they pleased. Feelings ran so high that Beckham finally brought a lawsuit against Potter for trespass in 1848,[11] claiming that he continued to depasture sheep there even though Lockyer had abandoned the run and had no licence. Potter argued, however, that the run had never been properly abandoned and that several hundred cattle and a number of sheep had been left there by Lockyer so that Beckham could not claim exclusive possession. The case was narrowly decided in favour of Potter by a three-quarters majority verdict.

THE POTTERS

Whilst Lockyer was the formal tenant at Cavan throughout this period, it was his son-in-law, George Potter, and daughter, Emily, who occupied Cavan. Emily Lockyer had married her husband, a retired army officer like her father, in 1836. From an inventory of items sent from Lockyersleigh to Cavan in February 1839,[12] it is clear that the young couple were already settled at Cavan and that Potter had turned to pastoral pursuits. Although a romantic and idyllic place on the river, Cavan was not without its dangers, being beyond the Nineteen Counties and more isolated from other settlers than grants closer to Sydney. Armed bushrangers roamed the countryside in this period, regularly attacking homes, stores and mail coaches, sometimes with fatal consequences for the victims.[13] However, there were some civilising influences in nearby Yass, including a Protestant church and a library with a subscription reading room established in 1839, both of which Potter was a founder member.[14] According

to Potter's second son, George Thomas, his parents were a highly respectable couple and very well off, boasting amongst their possessions: 10,000 sheep, two bullock teams, two drays, cattle, some fine horses and a phaeton.[15] Emily was reputed to be a most charming and cultured woman, who moved in the best military and gubernatorial circles prior to and during her marriage.

Fortunately, the affluence of the Potters does not appear to have been seriously compromised by the sudden collapse of pastoral expansion in 1840. About this time, the price of wool began to fall, cheap labour became in short supply with the end of transportation to New South Wales, and a three-year drought prevailed. Speculation had been rife, with land and stock being bought for sums far exceeding their real value, and when the crash came it was so significant that it subsequently ranked as one of the three great depressions of Australian history, along with those in the 1890s and 1930s. As the bottom fell out of the wool industry and land and stock were sold in a falling market to meet financial obligations, all levels of society were affected and ruin was widespread amongst traders, settlers and bankers alike. There were few pastoralists who escaped the correction; by 1843, a sheep could be worth as little as sixpence! Nothing like this had been seen in the colony before; one of Cavan's neighbours, Henry O'Brien, discovered[16] that the only thing to do was boil the flock down for tallow, as this yielded more income per head than a live sheep, even one in good condition.

In spite of these circumstances, the Potters carried on at Cavan and employed a number of servants, including a shepherd, Donald McKinnon. He was a Scotsman, sponsored by Lockyer at £25 per annum,[17] to emigrate to New South Wales where there was a shortage of skilled labour. Arriving on the *William Nicol* in 1837 with his wife, Mary, and their first child,

— *Mary McKinnon*

mercilessly teased by the other hands on the place who taught him English swear words which he then used without knowing their meaning![20] Subsequently purchasing land of their own, the McKinnons went on to have ten children, including Annie, who married Thomas Franklin. He was the uncle of Stella Miles Franklin, the author of several Australian classics, including *My Brilliant Career*. According to her biographer,[21] Miles Franklin based the character of Meg Syme in her novel *All that Swagger* on Annie McKinnon, declaring her to be her favourite aunt and immortalising her in Australian literature.

THE SNODDENS

After ten years of residence at Cavan, Emily Potter's position was to change dramatically through an ill-advised second marriage after her husband's death at Cavan on 20 October 1849. Three years before, he had suffered a serious riding accident that confined him to bed until he died.[22] Emily's father then took over responsibility, combining the operation at Cavan with his landholding at Lockyersleigh, which had increased to over 11,000 acres. According to Emily's obituary, 'handsome dividends were the result' of this joint enterprise and Emily was able 'to live in affluence' at Cavan with her eight children. Moreover, whilst Potter had died intestate, Emily was allowed to 'retain undisturbed possession, so as to have all the indicia of ownership'[23] of all Potter's personal property, and also to continue in possession of the farm at Cavan. As a widow, she wholly enjoyed these benefits for two years until her marriage in 1851 to Henry Snodden, the overseer and former employee of her late husband.

Snodden was dishonest and woefully abused his position as

Jessie, the McKinnons were no doubt glad to have escaped the poverty and starvation of the Scottish Highlands prevailing at the time. Leaving Sydney with a flock of sheep, they made their way to Lockyersleigh, where they stayed three months before moving on to Cavan.[18] Mary McKinnon travelled in a dray and reportedly became one of the first women to cross the ranges between Gunning and Yass.[19] Employed by Potter for more than six years, Donald McKinnon spoke only Gaelic and was the husband of a wealthy widow. From the large and convoluted file relating to Potter's intestacy, it is clear that Snodden, who somehow became the named tenant at Cavan and responsible for the payment of rent, sold all that he possibly could of Potter's possessions. He did this without accounting for them to his wife or to her late husband's estate, and deliberately ran down the Cavan stock and property in favour of building up his own separate flock.[24] History does not relate whether Lockyer had left at Cavan any of the Riley Saxons which he had purchased after William Riley's death or whether they had all been sold. If he had left any of them at Cavan for Potter to manage, it is probable that they were later either syphoned off by Snodden for his own ends or carelessly mixed up with inferior stock.

In any event, Snodden behaved so badly that Lockyer was induced to swear an affidavit on 27 February 1855 in which he attested that Snodden 'has lately very much ill-treated her [Emily] and has advertised the goods belonging to the Estate of G.T. Potter ... for sale. I believe that the said Henry Snodden after realising the said Estate will take possession of the proceeds.' Moreover, the person in Yass responsible for intestate estates, J. Stiles, wrote to his principal on 8 March 1855 that 'I believe the whole has been greatly mismanaged, for in the course of conversation, Snodden admitted that in one year lately, he had made from various sources £1,500 – but went on to say it had been all swallowed up.'

Not only did Snodden wrongfully sell most of the furniture at Cavan, but also a quantity of hay belonging to the estate which the purchasers later refused to surrender to the authorities on the grounds that Snodden owed them money. Snodden's defence for his conduct was that Potter owed him arrears of wages and the purchase price of a pair of horses that he had sold to Potter before he died. As regards the hay, Snodden maintained that he had grown this himself at his own cost and, as he was the tenant of the land from which the hay came, he was entitled to the proceeds. After more than a year of litigation, a small portion of the value of Potter's estate was recovered, mainly by the sale of unspecified sheep (many of which were said to be diseased), and Emily was granted a one-third share, the remainder being divided amongst her children.

— *Donald McKinnon*

93

Shortly before this squalid affair, a further incident took place that confirms Emily's misfortune. On 13 July 1853, one of the farm servants, a bullock driver called George Stephens, abducted her fourteen-year-old daughter, Susan, and the couple were found four days later in a nearby public house.[25] Stephens was sentenced to two years' imprisonment in Goulburn Gaol for unlawful abduction and the unscrupulous Snodden, Susan's stepfather, claimed expenses of £300 against Potter's estate! In spite of Snodden proving to be a most dishonest and untrustworthy husband, Emily appears to have remained with him at Cavan, moving with him to Tumut in 1860, and also bearing him more children in the meantime.

STUART DONALDSON

The architect of the leasing of Cavan and Narrangullen to Major Lockyer and his family had been Stuart Donaldson, administrator of William Riley's estate and joint executor under Honoria Riley's will. Most importantly, he was also guardian of the Riley orphans who were living with Honoria's sister at Denham Court. Invested with these positions of trust, Donaldson was not only responsible for the management of Cavan, but for all the Riley affairs in addition to the welfare of the Riley children. Although Donaldson was a joint executor of Honoria's estate together with Thomas Blomfield, there is no surviving documentation at all that suggests Blomfield took the slightest interest in these matters. On the contrary, Donaldson seems to have had a free hand, which gave him enormous power. Whether he exercised this honestly and fairly in relation to the Riley children is an open question, for within a few years of the eldest child and only son, Alexander Raby Riley, reaching

> Invested with these positions of trust, Donaldson was not only responsible for the management of Cavan, but for all the Riley affairs in addition to the welfare of the Riley children.
>
> —

his majority, there was little left of what should have been a spectacular inheritance. Donaldson's integrity and ability never appear to have been questioned by historians. He was to become a successful merchant and pastoralist as well as a notable public figure, knighted for his services in 1860. Long before all this celebrity status, however, and at the time of his appointment as administrator of William Riley's estate, he claimed that William owed him the not insignificant sum of £2,600.[26] How this debt was incurred and for what reasons remains a mystery and is not explained in the probate files. Did it have something to do with William's suicide? Had William lost the money pursuing a well-documented passion for horse-racing and had Donaldson picked up the tab? Was there some other cause more personal? Whatever the explanation, the debt may have been a useful spur to Donaldson to sort out the Riley affairs, and, certainly at this point, he appeared to be acting in good faith, applying himself to his duties with considerable energy and determination. For example, according to the voluminous file of correspondence with the Colonial Secretary amongst the Riley papers, he engaged in a lengthy dispute to have the quit rent on Narrangullen withdrawn for being far too onerous.

Quit rent was a controversial scheme that underwent various reforms in early land legislation in New South Wales, but essentially it was a hangover from the tenurial system of landholding in England, whereby anyone who acquired an estate from the Crown was regarded as its tenant and subject to perform certain services. In granting freehold estates in colonial territories, it was the practice of the Crown to reserve tenurial services and substitute them in the form of quit rents. These imposed a real burden on grantees and their successors in title, especially as they were often fixed at a rate that had insufficient regard to the quality of the land in question and its productivity. Ultimately, quit rents were reduced to a nominal sum only, but not in time to alleviate the burden on the Riley estate.

As Donaldson explained to the Governor, Sir Charles Fitzroy, he had leased Cavan and Narrangullen to Lockyer for 'a very moderate rental' of £75 per annum, whereas the annual quit rent due on the 10,000 acres at Narrangullen was £83.6s.8d.[27] Why a greater rent on these improved landholdings could not be achieved is unknown and unexplained in the correspondence. The rural depression of the 1840s had not yet begun and the price of wool was still strong so that properties such as Cavan and Narrangullen should have been relatively easy to let for sums sufficient to cover taxes and essential expenses. Possibly the inferior quality of the land compared to the Yass Plains (which was of concern to Dutton when he was choosing a selection with William on behalf of Alexander Riley) may have had a bearing on the rental price.

Whatever the cause, the deal struck by Donaldson was not only insufficient to cover the quit rent due, but there would be nothing left over for the education and maintenance of the Riley children or to pay Honoria's sister the £300 per annum promised to her. Clearly strapped for cash in spite of renting out the other Riley properties (including Raby for £215 per annum)[28] and notwithstanding income from the Riley flocks generally, Donaldson waged a vigorous campaign with the authorities for more than six years from June 1844 to obtain immunity from or at least a stay of the quit rents due on Narrangullen until Alexander Raby attained his majority in 1854. Pleading on behalf of the minor children, yet recognising there was no right in law whatsoever to claim such a concession, Donaldson shrewdly threatened to abandon Narrangullen to the Crown and apply instead for a £10 annual licence over the land – a much cheaper option than paying quit rent. Donaldson was so determined in his efforts to argue his case that the matter was taken all the way to Downing Street.

Whilst the issue was pending, Donaldson effectively achieved the very thing for which he had been asking, as no quit rent on Narrangullen was ever paid in this period. A compromise was finally reached in 1851, ordered by an exasperated Lord Grey of the Colonial Office in London, that the annual quit rent should be reduced by fifty percent and the arrears between 1846 and 1851 should only be payable once Alexander Raby had achieved his majority.[29] In spite of this enormous concession and considerable saving clearly due to Donaldson's skilful and unrelenting arguments, the Riley estate had already begun to be eroded by sales of sheep and loans, whether or not they were strictly necessary for the maintenance of the Riley children and the payment of Christiana's annuity. How much money from the rent received on the Riley properties found its way into Donaldson's pocket for managing the Riley estate and to pay off the loan of £2,600 owing to him on William Riley's death is not known. Possibly his remuneration and loan repayment came from other sources, including stock sales.

A very troubling aspect of Donaldson's involvement in the Riley affairs was his decision to move the Riley Saxons from Cavan, Narrangullen and Raby to his own properties.

—

A very troubling aspect of Donaldson's involvement in the Riley affairs was his decision to move the Riley Saxons from Cavan, Narrangullen and Raby to his own properties at Clifton and Tenterfield in northern New South Wales, on the border of Queensland. Although this could arguably be justified by the fact that Cavan, Narrangullen and Raby were leased (as well as the other lesser Riley landholdings), it potentially created a conflict of interest between Donaldson and his ward, Alexander Raby, who was heir to all the Riley pastoral properties and the Saxon sheep. When this valuable flock was dispersed at various stages between 1837 and 1859, some of the proceeds may have been applied to pay for the education and maintenance of the Riley children and some may have been used to repay the loan from Donaldson of £2,600, but these sums would have been relatively trifling compared to the total amount raised. Notwithstanding this, the sale proceeds appear to have had little or no impact on conserving or growing the Riley estate which, on the contrary, suffered an eventual collapse of epic proportions.

The first known sales of the Riley Saxons were only a year after William's death, a reaction no doubt to the leadership vacuum suddenly created in the Riley organisation, but possibly also to the debt owed to Donaldson by the estate. A small portion of the sheep was then sold in February 1840 in the Hunter district, on their way up to Donaldson's properties.[30] In all likelihood, this was necessitated by the combination of a dry summer and lack of feed on the stock route and the downturn in the pastoral economy which existed at that time. Harder to explain are the main sales of Riley Saxons in the mid-1850s that took place shortly after Alexander Raby reached his majority, the proceeds of which are impossible to trace. No reasonable bill for agistment on Clifton and Tenterfield could possibly have come close to the value of the Riley flock and cause such a sale. Nor could provision for the two remaining minor Riley children require such a measure. Moreover, why were the Saxons not returned to at least one of the Riley properties and Alexander Raby or a manager installed to oversee what should still have been a very profitable business?

According to a sale notice in the *Maitland Mercury* of 31 January 1855, 400 'very fine woolled well-bred RAMS, the progeny of imported German rams, selected in Silesia from the celebrated flocks of Prince Lichnowski [sic] and Baron Bartenstein' were available for purchase at Tenterfield. The following year, both at Tenterfield and at Clifton, a huge sale was advertised in the *Tenterfield Empire* on 4 February of over 11,000 sheep, including 800 rams bred from the same flocks as had been advertised the previous year. The stock was specifically stated as being 'from the well known flocks and herds of S.A. Donaldson Esq'. This was a clear misrepresentation, unless Donaldson had previously purchased the sheep from the Riley estate or had been given them (which is most unlikely) either for management services he had rendered or in repayment of debts. However, there is no evidence of any of these scenarios; on the contrary, subsequent events suggest that this may have been fraud or theft and part of a grander scheme by Donaldson to divest the Riley estate of all of its assets and feather his own

— *One of several mortgages between Alexander Raby Riley and Stuart Donaldson*

nest. What should have been an enormous bonanza for the Riley family seems simply to have disappeared.

A final sale of 1,600 ewes in January 1859[31] sealed the demise of the Riley stud. The dispersal of so many Saxon sheep in the 1850s would have raised a substantial sum running into tens of thousands of pounds and should strictly have been for Alexander Raby's account; yet it appears that by 1857, he was forced to begin borrowing large sums from Donaldson, starting with a mortgage over Raby for £4,000 which was later increased to £9,000.[32] Then, in 1858, Alexander Raby borrowed further sums from Donaldson secured over Cavan, Narrangullen, Warroo, Ousedale and Raby. Two mortgages have been found for £5,000 and £5,600 respectively, both at five percent interest and repayable in 1863.[33]

Why these loans were necessary and for what purpose they were made is unknown. No doubt the Riley estate had suffered during the pastoral depression in the early 1840s and may have incurred debts in that period, but it is almost inconceivable that fifteen years later the whole or greater part of the sale proceeds of the sheep would be required to pay these off, leaving Alexander Raby with little or nothing. Equally, even if Alexander Raby was a spendthrift (for which there is no evidence), could he really have gone through such an enormous fortune within two years of reaching his majority? The most plausible explanation seems to be that Alexander Raby was cheated by Donaldson or involved by him in some form of speculation or business arrangement that served to reduce Alexander Raby's circumstances whilst enhancing those of his guardian.

ALEXANDER RABY RILEY

Little is known of Alexander Raby's character, but there is no suggestion that he was anything other than a decent young man who was very much subject to the influence of his guardian. His father had died when he was not quite three years old, and he had grown up in the privileged and sheltered environment of Denham Court with his sisters and many cousins. His maternal grandfather, Richard Brooks, had died the same year as his father, William Riley, and his uncle, Henry Brooks, died when Alexander Raby was only eight. He therefore probably never had the benefit of much guidance from any senior male member of the family or of being rigorously groomed for business in the same way his father had been by Alexander Riley.

Moreover, Alexander Raby exhibited in his youth an academic rather than a practical bent, being one of the first cohort of students at Sydney University where, at the age of nineteen, he studied Chemistry and Experimental Physics, and was awarded the Chemistry Prize in 1853.[34] Nonetheless, it seems from university records that, for reasons unknown, he never completed his degree. Perhaps he was plagued by the same depressive illness that had afflicted his father and great uncle, or he found his studies dull and lacked the necessary motivation to finish them, or he was simply impatient to try his hand at farming and carry on the Riley family tradition. Whatever the truth, his fortunes became increasingly entangled with the affairs of his guardian, Stuart Donaldson.

In 1854, after his university studies and turning twenty-one, Alexander Raby went to live at Donaldson's adjoining stations of Clifton and Tenterfield, rather than residing at any of his own properties, which had all been leased.[35] At Tenterfield and Clifton, he became joint superintendent with Donaldson's

Alexander Raby exhibited in his youth an academic rather than a practical bent, being one of the first cohort of students at Sydney University where he was awarded the Chemistry Prize in 1853.

—

brother-in-law, John Cowper. The exact nature of the arrangement with Donaldson is not known, but according to a number of advertisements in the local newspapers, it seems that they also had a trading partnership for sheep and cattle, which was ultimately dissolved in January 1862.[36] Alexander Raby also opened a store at Tenterfield, but this, too, seems to have come within Donaldson's control. Through his firm, Donaldson Graham & Co, Donaldson became the store's major creditor to the tune of £2,348, having also sold Alexander Raby the land on which the store was built.[37] It is hard not to conclude from this and the loans secured over the Riley properties (possibly partly incurred to help finance the land and store at Tenterfield) that Donaldson was pursuing a deliberate strategy against his ward of financial dependency that would ultimately lead to Alexander Raby's ruin. Never having much grasp of trade or finance, the Riley heir was an easy target that Donaldson could probably not resist exploiting.

The final act in this tragedy was as cruel as it was prolonged. Whilst the majority of the Riley Saxons were sold two years after Alexander Raby turned twenty-one, he appears to have hung onto his inheritance of Raby, Cavan, Narrangullen, Warroo, Burwood, Ousedale and Malton for about ten years before he was forced to sell each one due to an insurmountable debt

burden. Since attaining his majority, Alexander Raby would have been in charge of these assets, at least nominally, but Donaldson had clearly enmeshed him in liabilities that could never be met from income. Donaldson, the main beneficiary of this scheme, took possession of the Riley flagship property of Raby on 7 October 1866, Alexander Raby being unable to repay the mortgage.[38] Moreover, both Cavan and Narrangullen were sold in 1862 for £1,425 and £2,000 respectively, significantly less than the value of the mortgages in Donaldson's favour.

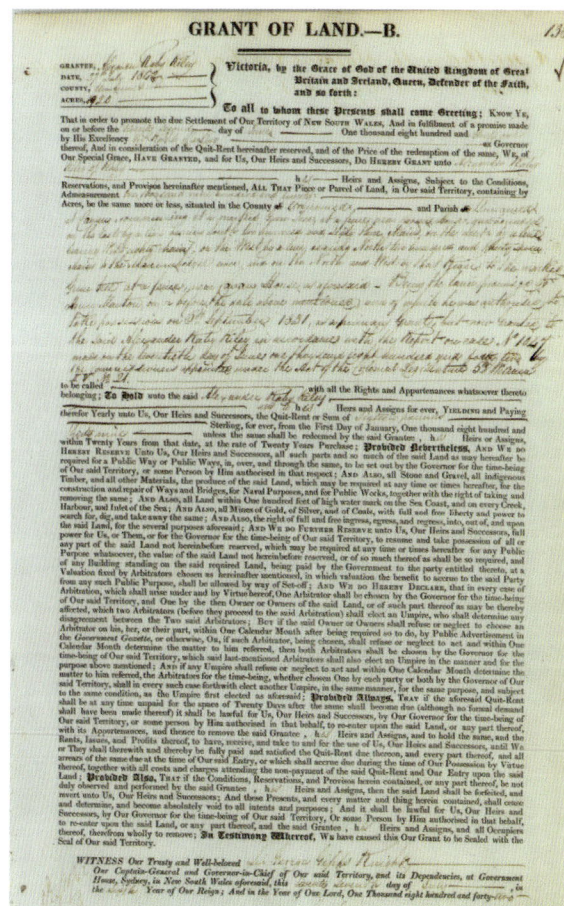

— *Formal grant of Cavan to Alexander Raby Riley, 1842*

> Could a man with no obvious vices and who was once worth so much be reduced to this by his own poor judgment and ill luck, or was he cheated by the person appointed to manage his affairs?

—

Warroo was also sold in the same year for an unknown sum, which would have been far less than Cavan or Narrangullen as it was a much smaller property.

Before these sales, however, Alexander Raby had been obliged to argue with the Colonial Office in London to have the second grant of 5,000 acres at Narrangullen completed.[39] The first grant of 5,000 acres had been formalised in 1842, but the government had been dragging its feet about the second grant, ostensibly on the grounds that the Rileys had not proved the additional land 'necessary for the preservation of the Saxon sheep', but in reality, because land available for selection was becoming in short supply and large grants of 10,000 acres were no longer looked upon favourably. In the end, the second grant was formalised in 1855, as Alexander Raby pointed out that the Crown had demanded quit rent on the land which it could not have done if title was disputed.

Just as the proceeds of sale of the Saxon sheep disappeared, the sale of each and every one of the Riley properties does not appear to have left Alexander Raby with anything more than a pittance. Somehow, he avoided being declared bankrupt, but it must have been a narrow escape. He had married Harriet Taylor in 1860 and lived with her and their children at Tenterfield, before moving in 1867 to a modest cottage in Rosedale,

Victoria, near his sister, Christiana. Something seems to have finally hit home when Alexander Raby was forced to close the Tenterfield store shortly before quitting the area and he tried at last to disengage himself from the Donaldson style of paternity. He not only surrendered his position as joint superintendent at Tenterfield and Clifton, but he also successfully applied in 1865 to have revoked Donaldson's Letters of Administration over William Riley's estate.[40]

But it was all too late and, ultimately, Alexander Raby died a pauper in Gippsland, Victoria, on 31 March 1881 with only £150 to his name, no real estate whatsoever and debts of £210.[41] The inventory of his personal effects is poignant, listing his only remaining possessions as two cows, three horses, a handful of shares in mining companies and an 'old' buggy and harness. Could a man with no obvious vices and who was once worth so much be reduced to this by his own poor judgment and ill luck, or was he cheated by the person appointed to manage his affairs? The evidence, albeit circumstantial, points to wrongdoing on the part of Donaldson, whose wealth and status appears to have increased in direct proportion to the diminution of Alexander Raby's personal fortune. The whole story has a distinctly Dickensian flavour, and although impossible to prove, the likelihood is that Alexander Raby was swindled lock, stock and barrel by his guardian. What had been a substantial legacy created by his father and grandfather, making Alexander Raby one of the wealthiest men in the colony, was completely destroyed by Donaldson's avarice and abuse of trust. However, not everything of this great patrimony was lost. At the time of Alexander Raby's death, the Riley Saxon bloodlines had continued thriving in other flocks and Cavan itself was already enjoying something of a renaissance under the auspices of the Castle family.

— *View over Cavan entrance*

CHAPTER SIX

THE CASTLE–CALVERT PARTNERSHIP

THE SALE OF CAVAN BY ALEXANDER RABY RILEY

ON 27 JUNE 1862, Alexander Raby sold Cavan to Joseph Frederick Castle for £1,425. The property was described in the conveyance as 'commencing at a marked gumtree, at a fence near Cavan House, and bounded on the East by a line bearing South two hundred and sixty three chains, on the South by a line bearing West eighty chains, on the West by a line bearing North two hundred and thirty seven chains to the Murrumbidgee river and on the North and West by that river to the marked gumtree, at a fence, near Cavan House, as aforesaid ...'[1] When compared with the description of the property in the original grant to Henry Manton, the boundary of Cavan had not changed since 1831.

Castle was already familiar with Cavan, having previously obtained, with business partner James Calvert, a three-year lease from Alexander Raby over both Cavan and Narrangullen on 1 May 1860.[2] The combined rent for these properties was £150 per annum, double what it had been when Lockyer was the tenant, but insufficient to relieve Alexander Raby of his growing financial problems. Before the lease expired, Alexander Raby was probably forced to sell Cavan – a fortuitous event for Castle and Calvert, who had firmly established themselves in the district through the lease of Cavan and a separate lease of most of its run (after the departure of James Fitzpatrick) in addition to a freehold portion of the run purchased for £713.[3]

Notwithstanding this partnership, the conveyance of the

Previous pages — *Original Cavan homestead site. The topography is clearly replicated in Louisa Atkinson's watercolour of the same view, page 118*

original grant at Cavan was to Castle alone. Importantly, the purchase did not include any stock; the auction notices that appeared in the newspapers[4] for the contemporaneous sale of Cavan, Narrangullen and Warroo, firstly in March 1862 (when the properties were passed in) and subsequently in June 1862 (when they were offered for sale without reserve), did not mention any sheep, the majority of the Riley Saxons having been moved by Donaldson to his properties up north in 1840 during Alexander Raby's minority.

What is interesting to speculate is whether Castle and Calvert were attracted to Cavan because of its prior association with the Rileys and the Saxon sheep. Both Alexander and William Riley were dead by the time Castle and Calvert arrived as immigrants in Australia, but unquestionably their legend as pioneers in the wool industry lived on. The association of Cavan with the Riley Saxons had been brief, but potent enough to influence subsequent owners and certainly sufficient to be forever linked in the minds of the public to the great achievements of Alexander Riley. There is no doubt there was a certain kudos attached to the ownership of Cavan as a result of its Riley connection and Castle and Calvert would not have been insensible to this. In time, they themselves were able to build on the reputation of Cavan as a fine wool-grazing property.

JAMES CALVERT'S BACKGROUND

Whilst it was Castle who appears to have provided the capital for the purchase of Cavan, it was Calvert who initially went to live there and manage the property. Born James Snowden Calvert on 13 July 1825 in Otley, Yorkshire, he was educated in the industrial towns of Liverpool, Manchester and Birmingham,

> James Calvert, fascinated with the native flora and fauna of his adopted country, arranged to become one of Leichhardt's exploring party.
>
> —

but appears also to have had some background in farming. Although his father is identified in the records as a leather manufacturer, Calvert, on his marriage certificate, also described him as a grazier. Leaving England for New South Wales with his older brother, William, the young men arrived as assisted immigrants in February 1842 on the *Sir Edward Paget*, their occupation described as farm labourers. This perhaps gave a misleading impression, bearing in mind their education and the fact that they had some prospects in Australia through their sister, Harriet Dawson.

Harriet was the second wife of Robert Dawson, formerly chief agent in New South Wales of the Australian Agricultural Company with which William Dutton had also been associated, before both men were sacked over the fiasco at Port Stephens. Dawson, after a period back in England during which he argued vigorously with the Colonial Office to absolve him from his mistakes, was granted land of his own in New South Wales in the upper Hunter region and a small property south of Newcastle called Redhead. Here, he settled with Harriet and their son, Joseph, and this move may have influenced the Calvert brothers to join them.

Travelling on the same ship as the Calverts was the explorer Ludwig Leichhardt, with whom the brothers became friendly. Leichhardt was later to invite James Calvert on his first expedition to the north coast of Australia via the unexplored territory from the Darling Downs in the northeast to the British settlement of Port Essington, east of present-day Darwin. At the end of 1842 Leichhardt visited Robert Dawson's property, Redhead, where the Calvert brothers were working, and it is probably at this time that James Calvert, fascinated with the native flora and fauna of his adopted country, arranged to become one of Leichhardt's exploring party.

CALVERT, THE EXPLORER

A great deal has been written about this expedition and the courageous part which James Calvert played in it. When he agreed to join the party, Calvert was only nineteen years old and would have had little real idea of the terrible hardships and dangers ahead, nor indeed that the expedition would take more than a year to complete.[5] The fact that he volunteered to go and at his personal expense – providing his own clothing, supplies and horses – suggests that Calvert had an adventurous spirit. He was also, according to Leichhardt's account, a very genial, good-natured, talkative young man, 'full of jokes and stories'.[6]

His brother-in-law, Robert Dawson, was also involved, being one of many landholders and merchants who supported the expedition with money and supplies in the hope that the explorers would find vast areas of pastoral land that they could then acquire through grants and leasing arrangements that prevailed at the time.

Whilst Calvert left no personal record of his experiences, it seems that his main interest on the journey was in botanical discoveries. Leichhardt mentions Calvert several times in his journal published after the expedition, usually in the context of his finding an interesting plant.[7] But Calvert's botanising did not shield him from danger. During the course of the expedition he came close to death on at least two occasions. The first was in January 1845, when he and Leichhardt became lost in the ranges of central Queensland for several days, almost dying of thirst and heat exhaustion. The second was in June the same year, when Calvert was speared five times during an attack on the party by Aborigines. He somehow recovered and, after another five months wandering around the Gulf of Carpentaria and Arnhem Land with Leichhardt and the remaining men, they reached their destination, Port Essington. On the expedition's eventual return to Sydney in 1846, the party, including Calvert, were treated as heroes.

Although now a celebrity, Calvert seems to have opted for a quiet life, declining an offer to go on Leichhardt's second expedition and returning to farming instead. Possibly, he went back to Redhead for a short time, but there is also evidence in the *New South Wales Government Gazette* that he was already in the Yass district by October 1847, though it has proved

impossible to ascertain his exact whereabouts. Whatever his precise movements, Calvert retained his interest in Leichhardt's activities, writing a letter over a decade later to *The Sydney Morning Herald* on 11 November 1858, 'cherishing hope' that Leichhardt, who had commenced a second expedition up north and been missing for ten years, was still alive. Curiously, this letter gives Cavan as Calvert's address although it was not until two years later that he and Castle acquired the leases of Cavan and Cavan run. It seems, therefore, that Calvert must previously have been employed by Alexander Raby at Cavan or squatted on a portion of the Cavan run, though no other evidence to support this has been found.

PARTNERSHIP WITH CASTLE

How Calvert and Castle met and came to form a business partnership is also not known, but the initial introduction may have been through the Dawson family. From a handful of remaining letters,[8] it seems that Castle was a close friend of Robert Dawson's eldest son by his first marriage, Robert Barrington. The two men were similar in age and had arrived in Sydney about the same time, possibly meeting in the Hunter Valley where they each had farming interests for a while. It is also conceivable that Joseph, the son of Robert and Harriet Dawson, attended one of the prestigious schools Castle owned and had established in Sydney – both of which Castle was headmaster – and that Castle met Calvert through this connection.

Both Calvert and Castle were educated, intelligent men, much admired by their contemporaries. In addition, they each had some farming experience, especially Calvert who had worked with his brother-in-law at Redhead and farmed in the

After another five months wandering around the Gulf of Carpentaria and Arnhem Land with Leichhardt and the remaining men, they reached their destination, Port Essington. On the expedition's eventual return to Sydney in 1846, the party, including Calvert, were treated as heroes.

—

Yass and Cavan districts prior to Castle's interest in the area. Whilst it was Castle who purchased Cavan, it was Calvert who brought many of the necessary qualities to the partnership to make the venture a success. He possessed in large measure enterprise, resourcefulness and resilience as evidenced by his triumphant expedition north with Leichhardt. By any yardstick, he was a safe pair of hands and an upright man, publicly acknowledged by the fact that he was made a magistrate in Yass in 1859.[9] He was a perfect choice of partner for Castle who, until 1870, was an absentee owner living in Sydney.

The relationship between Calvert and Castle was warm and friendly as evidenced by a small number of letters from Calvert to Castle that survive.[10] The letters give the clear impression that Calvert was very much 'hands on', doing much of the manual labour himself and even accompanying the wool to Sydney before returning to Cavan with supplies. On 11 December 1869, Calvert wrote to Castle from Cavan that, 'I have as usual been busy with hoe and plough in corn and [illegible] to try and keep down the weeds which are really a nuisance this year. I have just got it done in a rough manner and the two men leave tonight to go reaping.' Calvert explained in the same letter why there was a delay in bringing the wool: 'It was my intention some months ago to start the wool for Sydney, go down myself after it ... but the continual rain delayed the shearing and also gave us such an abundance of weeds and scarcity of hands.' Unfortunately, none of the correspondence or surviving documents of this era indicate the type of wool being grown at Cavan and, in particular, whether pure Saxon sheep were kept there. Given the diaspora of Saxons in the district after William Riley's death, however, it is likely that the wool was fine merino based in whole or in part on Saxon bloodlines.

What is apparent, however, is that the wool industry was a

— *James Calvert at Cavan, possibly overlooking original shearing shed on Mountain Creek Road*

challenge to the new partners. Calvert's frustrations are clear in his letters in which he laments the fluctuations in the wool price and the difficulties in reading the markets correctly so as to obtain the maximum prices for stock. However, Calvert found consolation in his love of botany and the many horticultural ventures that he undertook at Cavan. He appears to have planted a vineyard and winemaking became a feature of Cavan life. In a letter dated 14 March 1870, Calvert explained that although winemaking would not take place that year because more than half the grapes burst from late rain, there was some compensation in the weather: 'We have had more peaches than

— Watercolour and pencil drawing of James Calvert by Louisa Atkinson, painted between 1869 and 1872

I ever saw at Cavan and plenty of quinces and more mulberries than ever,' and whilst 'the corn became as dry as paper ... we shall have a fair crop also of potatoes'.

Calvert's pride in his fruit, vegetable and cereal crops is evident in his correspondence with Castle. His passion for botany seems never to have been far from the surface, and at Cavan he had ample opportunity for experimenting and cultivation. In 1862, at the first International Exhibition in London, Calvert won a bronze medal for a number of botanical specimens (regrettably unspecified) and a silver medal for an exhibit of vegetable fibres, proposed as materials for paper-making.[11] According to an article on 7 August 1869 in the *Goulburn Herald*, he also received a silver medal for similar botanical specimens that he provided to the International Exhibition in Paris that year.

In addition to plants, Calvert was interested in unusual farm animals and he became an active breeder of angora goats, following in the footsteps of Alexander Riley who was the first to import them into the colony. Calvert managed to trace the descendants of that flock to a settler in the Monaro district and purchased several does from him.[12] He also acquired a beautiful buck through the Acclimatisation Society of Melbourne of which his great friend, Ferdinand von Mueller, was a founder. Writing in *The Sydney Morning Herald* on 14 April 1871, Calvert explained that he found the most successful flock was bred by crossing the cashmere goat with the angora, producing a hardier animal with a heavier, silky fleece and a longer, finer staple. To what extent Calvert's diversification into angora goats was a commercial success is unknown.

Whilst he was a keen scientist, Calvert was also very practical. He was a conscientious housekeeper, watering the homestead garden and taking care of the pet lamb belonging to Castle's little daughter, Eliza. He also had plenty of company whilst at Cavan, since his brother, William, lived there for a time (purchasing a small block of land in the area)[13] as well as his sister, Harriet. She was probably keeping house for one or both brothers as she was by this time separated from her husband who had returned to England in 1862. Her son, Joseph, seems also to have been part of the Cavan workforce in this period. When he married Marion Smith in 1862, he described himself on the marriage certificate as 'Settler, Cavan'. How long he remained there and in what capacity is unknown, but he subsequently moved to Picton with his wife, and by 1882 was living at Glebe in Sydney.[14]

Although there are few letters left, those from Calvert are very detailed, suggesting he was a diligent correspondent, reporting regularly to Castle with both Cavan and regional news. In his letter of 11 December 1869, Calvert includes a reference to the new local MP, Michael Fitzpatrick, whom he noted, 'is well received by all our great Guns here, though a Catholic'. Such was the religious divide then that a Catholic MP was worthy of comment!

The most important event in the surviving correspondence recounted by Calvert was a major flood at Cavan, which he described vividly in his letter to Castle of 2 May 1870. He began, 'The first thing that I must tell you is about the flood though I daresay you will have received fuller particulars than I can give you through the Papers – all through April was wet more or less but did not cause us to apprehend a serious flood. On the 23rd it was heavier than usual. On the 24th it began to set in heavy and continuously all night, but Monday night 25th told us plain that a heavy flood must ensue.' Calvert went on to describe how the river rose four feet an hour and how he packed up everything of value into cases, putting them into the corn store ready 'at the last extremity' to take them away to the shearers' hut. He even made the boat ready, tying it 'fast to the peach tree at the corner of the grain store'. When the water reached sixteen feet past the mulberry tree, Calvert thought it was time to leave and prepared the horses; but suddenly, the river stopped rising and then by eleven o'clock began to fall. Calvert stayed up all night keeping watch with a carriage lamp and reported that the only loss suffered was 'a few panels of fencing'. This was exceptionally lucky, as in Yass the water was nine feet higher than the treacherous Gundagai flood of 1852 in which many people lost their lives or their homes and livestock.

— *Louisa Atkinson, c. 1865*

MARRIAGE TO LOUISA ATKINSON

On 11 March 1869, Calvert married the prominent writer and naturalist, Louisa Atkinson.[15] She was an independent woman of extraordinary talent and popularity, renowned for her literary abilities and artistic skills in natural history as well as more generally for the writing of fiction and the painting of landscapes

Above left — *Self portrait of Louisa Atkinson, undated*
Above right — *Watercolour of parrots by Louisa Atkinson*

Above left — *Drawing almost certainly of James Calvert by his wife Louisa Atkinson, undated*
Above right — *Watercolour of fern, by Louisa Atkinson*

— Louisa Atkinson's birthplace, Oldbury farmhouse near Berrima, by Hardy Wilson, 1914

and portraits. It was unmistakably a love match, with Louisa being a little older than was usual for new brides at the time and both parties having a large number of interests and friends in common. Although Louisa was afflicted with a serious heart condition that could slay her at any moment, she lived life to the full as far as she was able. Upon their marriage the couple lived at Cavan,[16] before moving closer to Louisa's birthplace of Sutton Forest, near Berrima, New South Wales. Nonetheless, whilst Louisa's association with Cavan was brief, she produced some important and illuminating articles about the natural history of the district and painted the earliest known depiction of the original Cavan homestead.

LOUISA ATKINSON'S BACKGROUND

Louisa's interest in natural science, particularly botany and zoology, stemmed from her childhood, specifically the influence of her mother and the sublime scenery in which she grew up.[17] Born on 25 February 1834 at Oldbury Farm, Sutton Forest, Louisa was the youngest daughter of James Atkinson, a free settler from Kent. After working in the Colonial Secretary's office in Sydney for a time, James Atkinson settled on the land and became an innovative and successful grazier. The homestead which he built at Oldbury in 1826 still stands today and ranks as one of the finest in the district. Like Alexander Riley, Atkinson was a firm believer in the importance of raising the quality of wool produced in the colony and toured Saxony in 1826 to study the Saxon merino.[18] This begs the question as to whether he knew Alexander Riley or his son, William, or perhaps even William Dutton, all of whom were at the vanguard of importing the Saxon merino into Australia about that time and moving in identical commercial circles. Evidence of such a connection has, however, not been found.

Unlike the Rileys, Atkinson concluded that Saxon sheep were probably unsuitable for such a place as New South Wales, with their delicate wool and constitution resulting (as he thought) from the unnatural practice of keeping them indoors. He may originally have undertaken the visit with the intention of buying stock, as he was a champion of New South Wales being ideal grazing country for growing wool, but he ultimately left Saxony empty-handed. Later well known for his book, *An Account of the State of Agriculture and Grazing in New South Wales*, which was published in 1826 and became an Australian classic on early farming methods, Atkinson died, probably from cancer, when Louisa was only eight weeks old.

His widow, Charlotte Waring, Louisa's mother, was a woman of considerable talent and backbone. A former governess, gifted artistically and a keen naturalist, she remarried two years later one of her late husband's friends, George Barton, who turned out to be psychotic. Fleeing his terrible temper and outbursts of violence, Charlotte took her children to Sydney for six years. As Louisa was a delicate child with a heart condition, Charlotte decided against sending her youngest daughter away to school, educating her privately at home. It was undoubtedly her mother's love of the native flora and fauna, her artistic aptitude and her literary skills[19] that shaped the mind and future career of the young Louisa.

Upon returning to Oldbury aged twelve years, Louisa absorbed a wide-ranging knowledge of plants, animals and birds found in the bushland around Berrima. She trained herself methodically in the skills of a natural scientist, learning to classify and dissect her specimens, as well as to sketch and write about them with what became characteristic accuracy and a delicacy of style. When she was in her twenties, Louisa moved with her mother to Kurrajong, a particularly beautiful rural area of New South Wales. In this richly forested region and in

spite of bouts of ill health, Louisa could indulge her passion for natural history to the full, diligently studying the habits of native animals and birds and collecting numerous botanical specimens. As Patricia Clarke notes in her outstanding biography of Louisa, 'Kurrajong was the scene of her greatest endeavours in plant collecting and in publicising the beauty of the Australian bush … For many hours she would ride over rock-strewn, precipitous tracks, along narrow ridges and through dense fern gullies, on botanising trips.'[20]

LOUISA ATKINSON, NATURALIST AND JOURNALIST

By the time Calvert married Louisa, she was thirty-five years old and had been working as a journalist as well as an amateur botanist for about fifteen years. Described by the eminent naturalist Reverend William Woolls as 'the most interesting of Australia's daughters',[21] Louisa was the author of a series of articles that ran at roughly monthly intervals for over ten years in *The Sydney Morning Herald*, from March 1860 to June 1870, under the title 'A Voice from the Country'. These articles, mainly concerned with the natural history, native flora and fauna and geology of the districts of New South Wales that Louisa visited or in which she lived, are remarkable for their minuteness of detail and brilliance.[22] They are also notable for being the first sustained local writing of their kind, for being the first series of articles written by a woman and published in a major Australian newspaper, and for their concerns about the environment.[23] They became a source of fascination for a Sydney audience starved of information about local phenomena and a passport for Louisa into scientific circles of the day.

Whilst Louisa's association with Cavan was brief, she produced some important and illuminating articles about the natural history of the district and painted the earliest known depiction of the original Cavan homestead.

—

> With Louisa the journalist as Calvert's wife, the thread linking Cavan to the newspaper industry through George Riley, Alexander's father, was further woven into Cavan's history.

—

Through articles such as these that appeared not only in *The Sydney Morning Herald*, but also in the monthly *Horticultural Magazine* and in the *Illustrated Sydney News* before its closure in 1855, Louisa became acquainted with the distinguished botanists Reverend William Woolls and Baron Ferdinand von Mueller.[24] Supplying them both with specimens on a regular basis from her frequent field trips, it is clear from correspondence that she was on easy terms with both of these great men. Her work was also tremendously admired by the geologist and Bishop of Sydney, William Clarke, and by the leading entomologist, William Macleay.[25] This was a time when the strangeness of Australian flora and fauna was a source of wonder to scientists all over the world and in addition to their own expeditions, amateurs like Louisa were relied on to provide more information.

It is also likely that it was her articles on natural history and related topics that brought Louisa and Calvert together. Exactly how they met is unknown, but Calvert was one of Louisa's avid readers[26] and it is possible that he contacted her, having read an article she wrote for *The Sydney Morning Herald* concerning the missing explorer, Ludwig Leichhardt. Louisa's interest in Leichhardt may originally have been aroused when she was eleven years old and living in Sydney with her mother

and siblings, at the time Calvert and his fellow explorers had returned from the first expedition to the north. In her article dated 1 May 1865, Louisa expressed the view that more should be done to try to find Leichhardt, seventeen years after his last known letter in which he stated that he aimed to cross the continent from east to west and explore Sturt's central desert. Louisa put forward her belief that there was no conclusive evidence that Leichhardt had perished, a view shared by Calvert, as is clear from his much earlier letter to the same newspaper.[27] Even if it was not through her journalism that the couple met, they were probably introduced by their mutual friends in scientific circles. Calvert, with his great interest in botany, was well acquainted with many of the people who were Louisa's friends and colleagues.

With Louisa the journalist as Calvert's wife, the thread linking Cavan to the newspaper industry through George Riley, Alexander's father, was further woven into Cavan's history. At least three of Louisa's articles were written whilst she lived at Cavan. The first entitled 'Among the Murrumbidgee Limestones' and dated 11 May 1870, described the landscape of high hills and unique geological features on and around the property and neighbouring Taemas. Writing that the 'limestone imparts a peculiar aspect to the scenery', Louisa spoke of the limestone ribbons that everywhere wind about the surface of the hillsides. She wrote, 'The limestone ... is found in strata-like walls, sometimes hardly thicker than a house slate or higher, sometimes in large blocks; acre after acre is thus lined and ridged – sometimes curving in a very line of beauty and of grace – sometimes stretching away like the rough stone walls of Northumberland or Yorkshire – so regularly, we are puzzled to believe there is no design in the masonry.' The huge, distinctive fold in the limestone strata, known locally now as 'the Shark's

Mouth' and situated on the Murrumbidgee towards the original Taemas Bridge, is identified by Louisa as being 'a curious arch' in the hillside.

Louisa clearly rejoiced in her new home and its surroundings. In the same article, she evoked a strong sense of Eden, becoming quite lyrical in her descriptions: 'There is a warm look in the soil, red and glowing; a vividness in the green of the grass ... a glow on the porphyry hills – a mossy, dark tint in the high oaks on the river banks, and above all, the gleam and glitter of the rapid stream, where fish and ornithorynchus [platypus] and hydromys [water rat] are sporting; and the roar and rush of white water, breaking over the reefs, is always on the breeze.'

It seems that there was an abundance of native animals at Cavan in that period and Louisa enjoyed observing their behaviour and habitats. In addition to platypus and water rat, she referred to rock opossums with thick, short, rusty fur; echidna and the rare musk duck, 'a peculiar looking bird, with habits of a diver. A pouch extending below the lower mandible to the throat, and a crest at the back of the head ... [with] plumage of grey and black.'

In a second article written whilst at Cavan, 'After Shells in the Limestone' dated 24 May 1870, Louisa devoted her piece to perhaps the most interesting aspect of the limestone: its fossils. She gives an account of a walk along the margins of the Murrumbidgee on the Taemas run and, in particular, a shale bank fifteen to twenty feet high which is unmistakably the wall of fossils now known as 'Shearsby's Wallpaper'. Louisa wrote that 'Here were the shells – not in ones and twos, but by bushels; sometimes in limestone, hard and blue; sometimes in the shale – from this we could pick the shells loose ... often white and pearly, like a shell new gathered from the beach – or of a delicate lilac, apparently the original colour of the shell.' Regretting her lack of

— *Superb blue wren, scarlet honey eater and pardalotes by Louisa Atkinson*

— *Elms on site of original Cavan homestead*

necessary to explain how the different limestone fossils came about, she was prodigiously talented and had even taught herself dissection and taxidermy.[29] In a third article entitled 'Climatic Influences on the Habits of Birds', which also appears to have been written whilst at Cavan and published in *The Sydney Morning Herald* on 16 June 1870, Louisa's skills in scientific examination were evident. For example, her analysis of a bird from the far north, driven south by the drought, appeared as follows: 'In two specimens which I have examined there were no nostrils. A long slit in the roof of the mouth communicates with the windpipe and it may be that the bird, when diving, inflates the pouch beneath the lower mandible and can remain under water until this receptacle is exhausted.' Several other references to dissections in the same article carried out on birds by the author attest to her proficiency and knowledge in this area.

knowledge of geology, Louisa found the sight of these vast fossil deposits 'perplexing' and also noted how they differed from the 'large grooved shells' found near the Yass River on Humewood, a property belonging to the explorer Hamilton Hume,[28] who was a friend of Louisa's husband. On Cavan itself, Louisa identified fewer shells but many corallines, tinted red, and fossils 'like huge mud oysters' and others 'like feelers and suckers of large Cephalopodi [cuttlefish]'. Remarking on the great difference in the specimens between those at Cavan and Taemas, Louisa speculated that 'perhaps, tides may have had something to do with this', but she could not say with any certainty.

Whilst Louisa lacked the scientific knowledge and training

LOUISA ATKINSON, NOVELIST

Upon her marriage to Calvert, not only was Louisa a widely respected naturalist and journalist, but she was a famous novelist too. As none of these novels were written whilst she lived at Cavan, nor contain any discernible influences from that time, a full discussion about them is beyond the scope of this book. However, no mention of Louisa in any context is complete without reference to her literary achievements. She became the first Australian-born woman to publish a novel in her native land and the first Australian female author to illustrate her own book. Her debut novel was *Gertrude, the Emigrant: A Tale of Colonial Life*, published in 1857, when Louisa was only twenty-three. Like all her novels, it has a strong moral tone and is distinctive for being a genuine story of life in the Australian bush with no attempt to anglicise the setting or the plot.[30] The harsh realities of pioneer life and the problems created by the mingling of convicts and emancipists with a community of free settlers were themes explored in *Gertrude* and returned to in Louisa's subsequent novels.

Publishing four further novels before her marriage, one in book form, *Cowanda, the Veteran's Grant*, and three as serials in newspapers, 'The Ground of the Carlillawarra Claimants', 'Myra' and 'Tom Hellicar's Children', Louisa was an established and successful author by the time romance came knocking at her door. Famous for her vivacity, amiability and special radiance,[31] she would have had plenty of opportunity to meet eligible men through her scientific work and social connections. The fact that she did not succumb to any of them before Calvert may partly be attributed to her strong, independent character, her devotion to her elderly mother and her own delicate health. It is perhaps ironic that in spite of this, the characters in her novels exhibit a firm belief in marriage and motherhood.

MARRIED LIFE AT CAVAN

Whilst a considerable amount has been written about Louisa, there is almost no record remaining of her life at Cavan. Clearly, she was living there with Calvert at the time of writing her articles on the local flora and fauna for *The Sydney Morning Herald*. It is also presumed that she went to live at Cavan soon or immediately after her wedding at Oldbury in March 1869. Calvert gave his residence on the marriage certificate as Cavan, Yass, and, in the absence of any other evidence, would in all probability have taken his new bride there to begin married life.

Aside from Louisa's articles on Cavan's natural history, there

She became the first Australian-born woman to publish a novel in her native land and the first Australian female author to illustrate her own book.

—

are only two other items that indicate her residence there. The first is a letter found amongst Calvert family papers.[32] This was from Louisa to her former maid, Mary Kelly, at Oldbury. It is dated 8 May 1868, though this must be a mistake as Louisa was not married until the following year and yet she signed the letter as Louisa Calvert. Although the letter does not give the address of the sender, she is clearly writing from her marital home. In the letter, Louisa advises that 'On the 17th of this month Mr Calvert and I go down to Oldbury which we shall reach on the 19th about 5 o'clock.' Louisa explained that they would be staying about a week and planned to go on to Swinton to see about buying a property there. She went on to refer to her health, explaining that 'the heart attacks are surely more than a week apart and sometimes seem bad – still I am much stronger and able to walk about a little'. Louisa's health must have been a source of great anxiety for her and Calvert, but it seems he took great care of his wife, as she also wrote in the same letter, 'Mr Calvert has as kind thoughtful little ways as a woman and I want for nothing.'

The second item evidencing Louisa's time at Cavan, and unquestionably the most valuable, is a very beautiful and delicate landscape in watercolour that is the earliest known depiction of the original Cavan homestead, washed away in the great flood of 1925. This picture, which is now in the possession of Susie

Castle-Roche, the great, great granddaughter of Joseph Castle, Calvert's business partner, was only identified and attributed to Louisa during the course of research for this book. Following the conventions of the eighteenth-century picturesque, the Cavan watercolour nonetheless has a local reality about it that makes the location easily identifiable – the sweeping bend on the Murrumbidgee just below a distinctive peak, timbered with kurrajongs, and rising up from the river flat.

As well as her other abilities, Louisa was clearly a gifted artist, illustrating her articles on natural history with pictures of animals, birds, insects, reptiles, shells, fossils and landscapes. Some of her works are in the Mitchell Library but, tragically, the bulk of her illustrations for two books on Australian natural history that she prepared for publishing by Ferdinand von Mueller in Germany, was lost or destroyed in the confusion following her sudden death. Louisa also painted portraits: an unnamed sketch in the Atkinson archive in the Mitchell Library of a middle-aged man with an oval face and beard is unmistakably of James Calvert (see page 111).[33]

DISSOLUTION OF THE PARTNERSHIP

The Calverts appear to have lived at Cavan for approximately fifteen months after their marriage, before moving back to Oldbury about June 1870, the date of the last of Louisa's articles on Cavan. Louisa's brother was at that time occupying the family home and, no doubt wishing to establish a place of their own, the Calverts purchased a house at Nattai near Berrima a short time later. Before they left Cavan, however, the successful partnership between Calvert and Castle, which had spanned more than a decade, came to an end.

The successful partnership
between Calvert and Castle, which
had spanned more than a decade,
came to an end.

—

Above — *View of original Cavan homestead before the flood of 1925*
Opposite — *Watercolour of Cavan homestead by Louisa Atkinson, c. 1870.*
This is an almost identical view to that in the photograph on pp. 102–103

According to a notice in *The Sydney Morning Herald* of 6 October 1869, the partnership between Castle and Calvert was dissolved on 1 July 1869. A receipt dated 24 September 1869 from Calvert confirms that Castle gave him £500 cash and two promissory notes of £500 and £635 respectively 'in full settlement of my co-partnership in Cavan Station'.[34] A later receipt dated January 1871[35] shows that the promissory notes were paid in full. No doubt the money that Calvert had tied up in the partnership was necessary to finance his new home with Louisa. It was probably characteristic of Calvert's personality that although the partnership was dissolved in July 1869, he and Louisa stayed on at Cavan for almost another year, to ensure a smooth transition to the new residents, the Castle family.

Calvert had contributed a great deal whilst at Cavan both by improving the property itself, but also as a stalwart member of the community. Aside from being a magistrate, he is reputed to have been the first person to propose a bridge over the Murrumbidgee at Taemas, northwest of Cavan,[36] which was only completed in 1887, long after Calvert had left the district. Nonetheless, the bridge was a hugely important development, as it finally provided a safe route to Yass for those living south of the river and enabled them to ship their produce direct to Sydney via the railway opened at Yass Junction in 1876.

DEATH

Once Calvert and Louisa moved back to Oldbury in mid-1870, Louisa wrote her last novel, *Tressa's Resolve*, and also began a new series of articles for the *Sydney Mail* under the heading 'My First Bush Home'. At around this time, Louisa was pregnant with a daughter, Louise Snowden Annie, to whom she gave birth on 10 April 1872. Only eighteen days later, after seeing Calvert's horse return home without its rider, Louisa died of shock,[37] 'cut down like a flower',[38] her weak heart unable to bear the strain of what might have happened to her husband and of the childbirth she had endured only a few days before. Calvert, who had merely fallen from his horse without injury, was broken-hearted. The man Leichhardt had once described as the life and soul of the party, became a virtual recluse. In spite of having a young daughter to care for, he never recovered from his wife's premature death. He died in Sydney in 1884, aged only fifty-seven.

Above left — *James Calvert, c. 1880*
Opposite — *Aerial view over Cavan and the new Taemas Bridge*

CHAPTER SEVEN

JOSEPH CASTLE

THE CASTLES ARRIVE AT CAVAN

THE CALVERTS' MARRIAGE IN 1869 that ended so tragically only three years later was the trigger for the arrival at Cavan of the Castle family. The Castle–Calvert partnership had been dissolved and Cavan was without a captain to steer it through the many vicissitudes of pastoral endeavour. Seizing the moment to retire from a high-profile and successful career in teaching, in 1870 Castle moved with his wife, Wilhelmina, and their twenty-one-year-old daughter, Eliza, from Sydney to Cavan, apparently without a backward glance. It would have been a big change for the two women, used to a degree of society and comfort in the urban surrounds of Sydney, but for Castle it may well have been the realisation of a dream. As his friend Robert Barrington Dawson wrote two years later, 'Your account of things at Cavan interested us greatly and you must not spare us any details pastoral, pectoral or agricultural. On the whole, I gather that the life bucolic suits you thoroughly after a fair trial of it and that Mrs C finds the bush by no means unendurable.'[1]

JOSEPH CASTLE'S ORIGINS, EMIGRATION AND FIRST MARRIAGE

Although Castle was born in Birmingham on 16 August 1811, he appears to have grown up fifty miles north in Nottingham, where his father was a lace manufacturer. Lace making had become an important industry in the East Midlands, which had started as a cottage industry and exploded in the early 1800s into machine-made manufacture, producing lace of world-class quality.[2] By 1832, there were over 180 lace makers listed as operating in Nottingham alone; with their excellent reputation,

Above — *Castle-Roche family graves on hill above new Cavan homestead*
Previous pages — *Driveway down to original Cavan homestead site*

lace makers had become very prosperous. Growing up first in Hockley and then in Mary Gate,[3] the centre of Nottingham's lace market, Castle must have been well educated and may even have been to university as he became a schoolmaster of some considerable repute. Regrettably, a search of the archives both in Nottingham and Birmingham (possibly where Castle's mother was born) has yielded no information about his parents or early life, save that his father died in 1832, a man of property and means.[4]

Castle was a beneficiary of his father's estate, but the amount he received is unknown. Before emigrating to New South Wales, Castle appears to have moved about 150 miles south to Bristol where he became a partner in and taught at Bedminster House Academy.[5] The school, however, closed with bad debts, and engaging himself to Eliza Goulstone, the eldest daughter of the Academy's proprietor, Castle departed England, without Eliza, arriving in Sydney in April 1838 on the *Orestes*. The reason for his emigration is unknown; he was one of eight children and the third of four sons so that his inheritance and prospects in England were probably limited in spite of his well-to-do background. One month after arriving in Sydney, Castle agreed to take over the management of Bellevue Hall School in Kent Street for four years, initially sharing a house with a merchant, William Drake, and his wife, Grace, whom he had met whilst on the voyage from England. Judging by his diary,[6] Castle was a very sensitive man and a devout Christian. In his first entry of 12 April 1838, he complains that, since his arrival in Sydney, he 'seldom enjoyed less of God's presence or was altogether more worldly', a refrain that occurs regularly throughout the early years of the diary.

Without his fiancée, Castle seems to have been very lonely and there are references to her in almost every diary entry.

He frequently calls for 'my precious darling, my cherub' to be restored to his arms and worries acutely about her health. On 7 October 1838, he wrote, 'my darling is unwell – Oh God – Great God Omnipotent I dare scarcely say my God save her – preserve her in health – then welcome poverty reproach and shame – yea all the ills of life … Oh God chasten me in any way but spare me my Eliza.' He would become extremely despondent if no letter from her arrived by the latest ship and descend into agonies of doubt about her wellbeing, sometimes fearing she was dead. Finally, after being apart for fifteen months and enduring a six-month period without receiving any news from her, Eliza confirmed that she was on her way to Sydney to join him. Castle's joy and relief were palpable, 'Oh! How can I believe it? What rapture! What a tumult of joy I feel. Great God! May I show faith my gratitude not only with my lips but eternally in life.'

Unhappily, Castle's longed-for reunion with his fiancée was short-lived. Whilst Eliza arrived safely on the *Arabian* on 26 June 1839, her health was frail. Although the couple were married a few weeks later on 13 July, she died on 10 August, leaving Castle desolate and in turmoil about his future. He lost interest in his work, being too sick at heart to continue teaching, and in early December advertised for a successor at the school. It was a pivotal moment when, full of uncertainty, Castle decided he would relinquish academia for farming. He wrote in his diary on 8 December 1839, 'This last week I have omitted to strive – I have not pressed on – I do not rise earlier – am not less slothful – when will this conflict cease. How much I am agitated in mind what to do? I fear to go farming – it may not be the will of God – … Lord increase my faith – and make my way plain before me.' Whether Castle had contemplated such a change of direction for any length of time is unclear as there are no earlier references in his diary to the idea. He does, however, appear to have previously spent several enjoyable school holidays on the land in the lower Hunter Valley and in the Illawarra region of New South Wales, together with his friends, the Drakes.

EARLY FARMING EXPERIENCES

In spite of serious misgivings, Castle purchased land at Hinton near Morpeth in the lower Hunter Valley in January 1840, struggling at first to find labour, ploughs and bullocks, and sleeping in a stringybark hut that he described as 'very smoky [sic]'. Disenchanted from the outset, Castle moaned on 17 May that 'my life is now so monotonous that I can scarcely record anything but that I sometimes dine out on Sunday unless I mention how many acres I have ploughed. My spiritual state I cannot speak of – for alas! I am dead – dead to mercy and warnings I fear.' It was an enormous change of lifestyle for a gentleman schoolmaster, and an isolated one without a wife. The importance Castle attached to intellectual stimulation is perhaps best evidenced by a diary entry of 3 July about an acquaintance, Mrs Pearse, but which also shows his growing humility: 'How kind this woman is to me – formerly I only esteemed people for literary attainment or some excellence bordering on Education and accomplishments. I am learning to form a truer estimate of Character for I love Mrs Pearse though ignorant for her truly affectionate heart.'

But farming did not go well for Castle right from the beginning. Rain in July ruined his first wheat harvest and he lamented that 'coming up the Hunter was one of the greatest pieces of folly I ever committed'. Nevertheless, by September, he admitted that 'In spite of annoyances I have been happier this

last week – enjoyed calmness long time unfelt.' He soldiered on in spite of mounting debts caused by drought, and dined regularly with his neighbours as well as making trips to Sydney to help relieve his boredom and loneliness. By January 1844, Castle was poorer than he had ever been, recording that 'All this month I have been harassed with accounts from Sydney which I cannot pay.' In August, he wrote, 'Times are such on the Hunter that a Lethean apathy pervades almost everything – so that I am almost tempted to say: Happy are they whom the Insolvent court has released from the struggle.'

SECOND MARRIAGE

Fortunately for Castle, he had his teaching skills to fall back on and, by May 1845, he was looking for school premises. Renting Wallalong House at Hinton for six months from ship's surgeon Dr Walter Scott, Castle opened a school in August with just three pupils: Frank, George and William White. In December, he accompanied the young boys home to Muswellbrook where he met the Whites' governess, Wilhelmina Roche. Proposing to her two weeks later on 28 December, Wilhelmina accepted.

The timing of the meeting was perfect, as a 'complication' in Castle's life had come to an end just five months earlier. Whilst memories of Eliza had never been far from his mind when alone on his farm (especially on the anniversaries of her arrival in Sydney, their wedding day and her death), and even though Castle had corresponded regularly with her family back in Bristol, he appears in this period to have become very close to Grace Drake, whose husband was a drinker and prone to idleness. When Grace died on 5 July 1845, Castle was privately devastated and, in a most unusual arrangement, Grace was placed in the same tomb as his late wife, Eliza. Clearly, had she survived, Castle's association with Grace could not, without huge scandal, have gone any further, and her death freed Castle emotionally to marry someone else.

On 14 April 1846, Castle and Wilhelmina Roche were married at Muswellbrook. Taking a house at Austenham, three miles from Sydney, Castle opened a school there in July for young gentlemen, starting with fifteen pupils. A Miss Lambie came to live with the new couple as housekeeper and life began to take on a steady rhythm. The school thrived, with boarders paying twelve guineas a quarter and day pupils four guineas with no extra charges of any kind.[7] Castle took on an assistant master and found someone else to visit twice a week to teach French. It is not known when Castle finally sold his farm at Hinton, but through the success of the school he was able in August 1848 to pay off the remainder of his debts arising from his farming days.

WILHELMINA CASTLE'S ORIGINS

Whether Wilhelmina (or 'Willie' as Castle was fond of calling her) ever took part in the running of the school is unknown. She was an Irish Protestant and a native of Dungarvan, County Waterford, who emigrated from Cork in June 1840 with her brother, Frederick Roche, on the *Elphingstone*. Almost nothing is known about her, other than that she came from a large family who claimed to be distantly related to the Irish peer, Lord Fermoy. On her departure for New South Wales, Wilhelmina left behind a broken-hearted cousin, Philip Dowe, who wrote seven pages of mournful, valedictory verses bidding her farewell and cherishing the hope of meeting her again in heaven. Amongst these verses, he wrote the lines:

Our fates are one if we are one in Him: our lives may run
In different channels, but they onward tend
To heaven their great attraction, and their end.
Like streams which travel to the self-same sea,
We seek the ocean of eternity;
And there, when earth's long wanderings are past,
shall mingle peacefully and rest at last.[8]

If Wilhelmina returned her cousin's love, she must have been a young woman of exceptional independence and spirit to leave him behind for an unknown destiny in a strange country so far away. According to the memoirs of Albert Piddington, a former pupil of Castle's who became a High Court judge, Wilhelmina developed in middle age into a 'masterful lady' and a 'maitresse femme', who was disciplined and invariably kind 'beneath a surface of severity'.[9] She was in the habit of prowling around the Cavan homestead 'prying into every one's doings and scolding the maids till her corkscrew curls, shaken in wrath and menace, transformed her fine head into a Gorgon's, bristling with as many snakes as the Medusa's.[10] Wilhelmina was clearly no fool and Piddington alleged further in his memoirs that 'it was her vigilance and firmness alone that had saved [Castle] and the station from the clutches of a designing trickster'.[11] To what episode Piddington was referring is not known, but it seems that Wilhelmina was perhaps more astute in matters of business than her husband.

— *The Castle family, c. 1855*

FAMILY

AS CASTLE'S WIFE, Wilhelmina tried her best to bear him children. According to her husband's death certificate, Wilhelmina lost eight male children in total, but the dates of only four male births have been found. Wilhelmina's first child was lost on 16 January 1847, after she had 'foolishly driven to town in a rickety cab & was so jolted as to risk her own life and lose that of her infant'. She then lost a pair of twin boys in 1848, before finally providing Castle on 6 October 1849 with a daughter who survived into adulthood, Eliza Grace Lambie. A son, William Frederick Fletcher, was born on 4 February 1852, but he died two years later. Eliza, who was named after three women – Castle's first wife, Eliza Goulstone; his 'friend', Grace Drake; and the faithful housekeeper, Miss Lambie – became the apple of Castle's eye.

CALDER HOUSE SCHOOL

Sometime in January 1856,[12] Castle moved his family and school to Calder House in Newtown near Redfern, a suburb of Sydney. The house had been built in the 1820s by James Chisholm, a migrant from Scotland, who had received a sixty-acre grant of land after serving in the New South Wales Corps.[13] Castle converted the house into a school which became one of the most famous and highly regarded private boarding schools for boys in New South Wales. According to a former pupil, Alfred Campbell, the school was 'most ably conducted' by Castle, who was himself a 'genial, kind-hearted gentleman' and a great conversationalist.[14] He was also an excellent, inspirational teacher, especially of Latin. Albert Piddington declared that instead of 'mere memory labour of declensions and terminations', he 'made Latin a living language'.[15]

Castle seems to have had a love of teaching that stemmed from his childhood, after he was forced to lie in a darkened room for three years following an accident with gunpowder that threatened his sight. During that period his grandmother, Lady Caroline Fletcher,[16] taught him Greek and Hebrew, and it was her patience and dedication that showed him the purpose and rewards in teaching. Castle not only inspired his pupils in the same way, his sympathy for the young endeared them to him. He was very well read and 'lived in his reading',[17] but he was also known to make his own quill pens and wrote in a beautiful copperplate. His elegant signature on the conveyance of Cavan bears witness to this skill.

Above — *Calder House school, with Joseph and Wilhelmina Castle at far right, c. 1860*
Above right — *Cutting from the* The Sydney Morning Herald, *20 April 1921*

PARTNERSHIP WITH FREDERICK ROCHE

In spite of the evident success of Calder House School and notwithstanding the failure of his farm at Hinton, Castle always remained interested in the land. In the late 1840s, he seems to have gone into some sort of business partnership with Wilhelmina's brother, Frederick Roche, who took up land on the Dawson River in the Leichhardt district of Queensland shortly after his arrival from Ireland. Roche's property was called Marydale and he lived there with his first wife, Helen, the daughter of a clergyman, whom he married on 28 July 1853. Roche and Castle kept up a regular correspondence, though only the letters from Roche survive.[18] They are full of information about wool orders, the problems and cost of labour and the need for supplies, as well as general complaints about squatting life. Many of the letters read as if Roche is accountable to Castle, detailing amounts drawn at the bank or sometimes enclosing cheques and invoices.

There is no doubt that Roche worked very hard and conscientiously to build up the business. On 19 December 1852, he wrote, 'I am not willing to buy more sheep as I can scarcely get the labour to look after what I have – also none are willing to stay in the district except for high prices – it is better to go on steadily than to be paying out money upon an uncertainty of the labour market ... I am worked to death to try and keep things right ... the sun has not caught me in bed for many a month.' By this time, he already had 20,000 sheep spread over several properties in the area; although the shortage of labour was generally widespread, an incident earlier in the year at Marydale had discouraged men from living in the district altogether. The occasion was the brutal murder by Aborigines of Roche's young nephew, William Clerke, on 7 April 1852.

Clerke had emigrated from Ireland in 1849 and been employed at Marydale as superintendent. According to the *Moreton Bay Courier* of 24 April 1852, the Aborigines were taking revenge on stockholders who had recently withdrawn their cattle from grazing on the river where, by means of theft, the animals had provided additional food for the native inhabitants. Attacking anyone who was in charge, the Aborigines killed Clerke, severing his head from his body with a tomahawk. His uncle Roche was distraught, writing to Castle on the day of the murder, 'I wish that I were a woman I could cry ... Please write to his father – I cannot for it has broken my heart – he was saving up £100 to bring his family over.'

Gradually, during the eighteen months after this tragedy, Roche grew defeated by the lack of a proper workforce and he wrote sadly to Castle on 8 November 1853, 'My dear fellow, it is time to quit this place.' However, it was not until March 1855 that Roche sold his 16,000 acres at Marydale plus a further 22,000 acres on Maud Creek. He wrote excitedly to Castle on 28 March, 'All my neighbours say it is a first rate sale and are all envying my good fortune in renouncing squatting for they are all like myself heartily sick of it.' It appears that Roche and his wife then went to Dalby in Queensland where he became a shopkeeper, district registrar and the first mayor in 1868.[19] They had four sons, one of whom, Frederick William, was born at Austenham House in April 1854, four years after Eliza Castle.

THE PURCHASE OF CAVAN

The sale of Marydale left Castle without a rural interest, and just two years afterwards he and James Calvert began making enquiries about purchasing Crown land on the Cavan run. It was not until the expiry of James Fitzpatrick's lease in 1860 that they obtained a small freehold portion of the run, a lease over the remainder and a lease over Cavan itself, the latter from Alexander Raby Riley. The actual freehold purchase of Cavan by Castle was not until 1862, by which time the idea of becoming a grazier in exchange for life in the city must have taken hold.

Nonetheless, perhaps feeling cautious after the experiences of Hinton and Marydale, Castle seems initially to have been content to stay teaching in Sydney whilst allowing Calvert to develop their pastoral interests at Cavan by himself. Calder House was prospering and continued to do so throughout the 1860s, judging by the various advertisements in the newspapers for tutors and term dates.[20] Castle, however, appears to have been a regular visitor to Cavan when he could get away from his responsibilities in Sydney. He seemed to appreciate the bush on these occasions and wrote vividly and affectionately to his young daughter, Eliza, on whom he doted. One of these letters,[21]

The actual freehold purchase of Cavan by Castle was not until 1862, by which time the idea of becoming a grazier in exchange for life in the city must have taken hold.

—

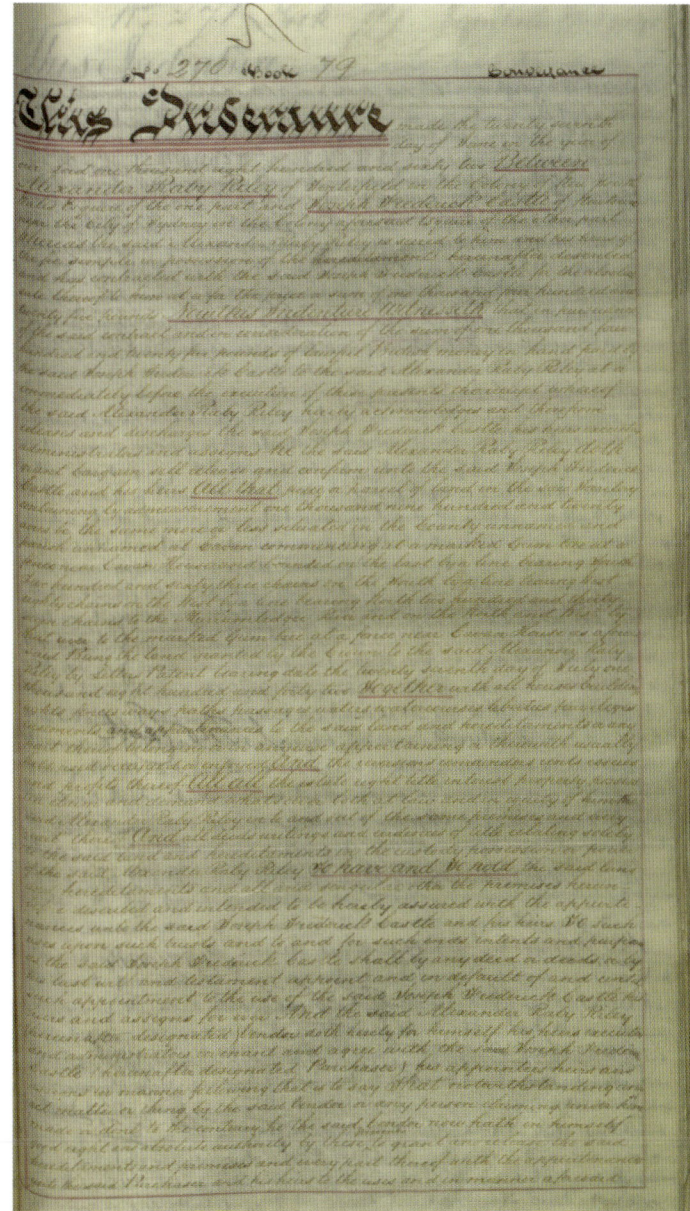

— Conveyance of Cavan from Alexander Raby Riley to Joseph Castle, 1862

which is undated but probably written in the early 1860s, reads as follows:

My darling Eliza,

I am sitting in Mr Calvert's nice little cottage on the banks of the Murrumbidgee, the largest river in Australia, you must look it out in the map and Henry Hall has just caught some fish for tomorrow's breakfast. I went to dine with Mr Hall – he lives seven miles from this and he has five such nice little girls. Two or three of them will be in Sydney and then I will introduce them to you for they have promised to come and see you. On our journey sometimes, I rode on horseback and managed very nicely. The horses were a little tired when they got to the station, but they have recovered now. I have not ridden Dandie since I came down so that he may be strong for me when I come back which will be soon after you receive this for before you read this note I hope I shall be halfway home. The station is very pretty plenty of grass and such beautiful clear water, you may see all the stones at the bottom of the water and sometimes little fish swimming. And then you can see such numbers of beautiful little calves playing with their mothers so quiet because nobody ever drives them about and a lot of horses and mares too some of these have little foals. I hope you have been a good girl and that you have given Mamma no trouble from ill temper or self-will for remember 'Christ pleased not himself' though he is Lord of all. Night and morning and often during the day I beg God to bless my darling child and to make her one of the lambs of Christ's flock [illegible hereafter].

It is not known how much time Castle and his family spent at Cavan in the ten years that he was an absentee landlord. Possibly they went there in the school holidays so that Eliza could benefit from the fresh air and freedom as well as learn firsthand a little more about nature. Even so, it was a considerable change of lifestyle for the whole family when Castle handed over Calder House School to a long-standing staff member, Mr Sly,[22] and they moved permanently to Cavan. For Castle, who always took a close interest in the management of the property as is clear from his correspondence with Calvert, it must have been an exciting time and it is easy to imagine that he threw himself into his new life with some relish. The reaction of his wife and daughter, now a young woman, to being transplanted in this way, is uncertain since no diaries or letters remain revealing their thoughts on the matter.

THE CAVAN DREAM

On the other hand, Castle left plenty of personal material to posterity. His enthusiasm for Cavan is perhaps best illustrated by the meticulous station diaries (numbering eleven volumes of minute writing) that he kept from 1872 right through to his death in 1883.[23] These diaries give a good insight into the rhythm of life at Cavan and the things that Castle was particularly interested in, as well as the productivity of the place. The property was essentially a wool-producing enterprise, and the amount of activity at Cavan increased notably towards the last quarter of every year as preparations were made for shearing, which commenced in October and often did not finish until mid or late December.

Regrettably, the diaries do not indicate what type of sheep were kept at Cavan in this period, but the influence of the Saxon merino is likely to have been prominent as it was in so many flocks in New South Wales by this time. In 1873, twenty-five bales of wool were produced at Cavan from 2,648 fleeces averaging 2 pounds 10 ounces per fleece. This tally increased steadily in the next few years so that in 1876 thirty-two bales of wool and 3,000 fleeces were recorded. By 1877, the wool clip amounted to forty-one bales; but this is the last year in which Castle entered such details in his diaries, possibly keeping these records in a separate register, since lost. In any event, given the type of country at Cavan, he was fortunate in his decision to concentrate on sheep, especially when the weather was dry. Writing on 24 March 1881 to his new sister-in-law, Kitty, Frederick Roche's second wife, Castle said, 'We are without rain and never in my long experience of Cavan did I ever see the place so dreadfully burned up. The sheep do well in spite of drought as there is always plenty of running water and I congratulate myself in having sheep instead of cattle (of which I have now only a few hundred) as sheep can be paddocked but in such a drought the cattle could not have been kept within bounds in a limestone country which burns up sooner than any other.'

— *Crossing the Murrumbidgee at Cavan with sheep, c. 1890*

Of particular note in the diaries are Castle's many references to his horticultural pursuits, an interest he shared with Calvert, especially the vineyard of which he seemed very proud. This vineyard was planted some time before the Castle family moved to Cavan and as early as December 1866, Castle wrote to his daughter, Eliza, 'Did I tell you that here were plenty of grapes at Cavan ... I have recommended Calvert to make some wine.' In the diaries, there are frequent entries concerning the maintenance of the vines, as well as the planting of vegetables (including asparagus, beans, cauliflower, carrots, radishes, potatoes, cabbages and lettuces) and the growing of fruit and nut trees (such as peach, apricot, mulberry, walnut and almond). The homestead would have been almost self-sufficient, as dairy cows for milk were also kept, as were pigs for bacon. Little seems to have occurred to ruffle the lives of the inhabitants in the Castle years. There were no major fires in spite of some very dry summers to which Castle refers, and the earthquake that suddenly occurred on 3 August 1876, 'a rumbling shock from west to east', did nothing more than shake the roofs of the buildings and make an ominous sound that rolled through the hills.

— *Ancient olive tree in old garden of original Cavan homestead*

JOHN THATCHER, A TRUSTED EMPLOYEE

Many of Castle's diary entries and several of Calvert's letters mention an employee called John Thatcher who appears to have been a key member of the Cavan workforce for many years, employed there initially as a stockman but later becoming superintendent and then overseer. Thatcher lived in the Yass district for over seventy years and, as acknowledged in his obituary, was 'always held in the highest esteem'.[24] Although his origins are obscure, it is known that his mother was a married convict, Mary Ann Holland, who was transported for seven years to Australia for stealing four silver spoons and various other household items whilst employed as a maid. The exact

date and place of Thatcher's birth and the name of his father are unclear but, from immigration records, it is most likely he was born on the convict ship the *Planter*, which brought his mother to New South Wales in 1839 at the age of twenty-seven.[25]

This convict 'stain' may have been the reason why Thatcher was always vague with his own family about his birth. After his death, colourful stories circulated regarding his paternity. One theory postulated that he was the son of Hamilton Hume who, after exploring the area in the mid-1820s, had settled there with his wife on the other side of the Murrumbidgee not far from Cavan. However, there is no evidence whatsoever to support this rumour, which may have gained currency because of Hume's own childless marriage. What is certain is that about five years after Thatcher was born, his mother contracted a bigamous marriage in 1845 to his stepfather, Thomas Thatcher, also a convict, with whom she had at least three more children. Thomas Thatcher appears to have had a positive influence on his stepson's life, teaching him the secrets of bushcraft and the skills essential to becoming a stockman.

Thatcher's obituary asserts that he started working at Cavan about 1852 at the age of twelve. Whether or not this is true, he was certainly there as a stockman by 1866. An account in *The Wallaroo Times and Mining Journal* of 2 May 1866 entitled 'Sticking up of Cavan Station' mentions a stockman by the name of Thatcher being one of several Cavan residents – including James and William Calvert and their sister, Harriet Dawson – who were taken hostage at gunpoint at four o'clock in the afternoon by two bushrangers that 'ransacked every room in the dwellinghouse, breaking open, with few exceptions, every article of furniture that was locked' and procuring some valuable firearms, gold and silver watches and two horses. Fortunately, no one was harmed and, after eating supper prepared by one of the servants, the bushrangers left with their booty about ten o'clock.

In 1867, Thatcher married Harriet Gledhill from Yorkshire. The birth certificates of their children confirm that Thatcher was a stockman at Cavan until 1874, whereupon he became superintendent of sheep and then, in 1876, overseer. Shortly before this last promotion, his half-sister, Eleanor, married William Calvert who was nearly twice her age, on 16 November 1874, helping to raise the Thatcher family above their humble convict origins. William is first recorded as being in the Cavan area in 1862, when he purchased a block of forty acres[26] on Mountain Creek at the southern end of the Cavan run and adjacent to a block of 160 acres in the joint names of his brother, James, and Joseph Castle. He named his land The Brook, which was recorded in the Electoral Roll for Yass Plains in 1870. Whether William was farming on this modest block is unknown, but he does not appear to have had the success in life that his brother enjoyed. According to family tradition,[27] he liked to drink, though he was nonetheless elected a sheep director for the district of Yass and appointed a magistrate. Around the time of his marriage, he seems to have sold The Brook to his brother-in-law, Thatcher, who owned a number of surrounding blocks, and moved to Picton with Eleanor where two of their three children were born.

In spite of now being in a senior role and no doubt an important person on whom Castle could rely, Thatcher had ambitions beyond Cavan and appears to have left his employment there about 1882 to begin a carrying business between Yass and Sydney with a team of ten bullocks.[28] Carrying could be a reliable and lucrative business for the hardy worker, as transport for the distribution of wool and other primary agricultural produce was in strong demand. Thatcher still maintained close ties with Cavan, and in the early 1880s he is mentioned several times in the Cavan diaries, working as a contractor during shearing or delivering goods with his dray. By the late 1890s, he had become a grazier himself, according to the description of his occupation on the marriage certificate of his second son, William (known as 'Tom'), and had purchased at least a further 300 acres of choice country between Mountain and Razorback Creeks.[29] Blocks with creek or river frontage were highly prized by all settlers, having the obvious advantage of being able to water stock easily without the need of a pump.

By the time of Thatcher's death in 1925 at the age of eighty-four years, The Brook comprised at least 500 acres and possibly as much as 940 acres of beautiful creek flats.[30] It was regarded as a 'fine property', occupied by his youngest son, John (known as 'Jacko'), who together with his elder brothers, Tom and Alfred, became notorious for running brumbies down from the nearby mountains and breaking them in for sale.[31] Through hard work and dedication, Thatcher had done very well for himself despite his inauspicious beginnings, leaving behind a legacy that his children were fortunate to inherit.

PRESERVING THE CAVAN ESTATE

The Castle era at Cavan appears to have been one of consolidation rather than innovation. During the Castle–Calvert partnership, vitally important land reforms came into force. However, the potential of these reforms to reduce the size of the Cavan run and disrupt the grazing practices of the whole Cavan enterprise really only emerged in the late 1860s, shortly before the partnership was dissolved and the lease on the Cavan run expired. It therefore fell mainly to Castle to take the initiative of keeping the property intact and new settlers at bay.

— Eliza Grace Lambie Castle, c. 1870

In 1861, land legislation in New South Wales underwent a complete revision to 'unlock the lands', ending the monopoly of a few over vast tracts of Crown territory and enabling more people to own a small property on which to make a living. The legislation was deeply unpopular with wealthier settlers, some of whom were occupying as much as 50,000 acres under lease, as indeed was Castle. Nonetheless, this principle of equality of opportunity regarding access to land was enshrined in the *Crown Lands Alienation Act* and the *Crown Lands Occupation Act* (also known as the Robertson Land Acts after Sir John Robertson, Minister for Lands and champion of the legislation).

Under the new legislation, any person (called a 'selector') could purchase between 40 and 320 acres (later increased to 640 acres) of unimproved and unsurveyed Crown land (excluding land earmarked for townships or proclaimed a mining or water reserve) at £1 per acre, provided he deposited twenty-five percent of the purchase price immediately and either paid the balance within three years (thereby acquiring freehold title), or if unable to do so, paid interest at five percent on the outstanding balance until the full amount was discharged. This scheme was known as conditional purchase and was dependent on improvements being made by the selector and evidence of residency. Additional adjoining land up to three times the amount of the original selection could also be leased, but this was often impossible if the surrounding land had been selected by another person or was already part of an existing freehold or leasehold title.

As is clear from entries in the Cavan diaries and on the Parish Map of Taemas 1884, the earliest map of the Cavan run available, Castle made a number of selections or conditional purchases in order to protect the Cavan estate from being broken up by the tide of new selectors. The exact amount of land that Castle acquired in this manner is unknown and cannot be accurately ascertained as the map gives insufficient details. However, it could have been up to 3,570 acres judging by a diary entry in 1888, five years after Castle's death, which records this as being the acreage then held by his daughter, Eliza, and her husband under the conditional purchase scheme. The presumption is that, before he died, Castle may have been unable to pay off some or all of these conditional purchases, and that these were subsequently taken over by Eliza and her husband and acquired by them as freehold once the remaining debt was satisfied.

As the Robertson Land Acts stipulated that a person could only hold up to 320 acres at any one time until he had paid off his selection in its entirety, the practice of 'dummying' or standing in for one's employer was common (until it was made illegal in 1875) and enabled a much larger amount of land to be held by a grazier. Thus, it seems that the faithful Thatcher agreed to act as a 'dummy' for Castle, purchasing under the conditional purchase scheme two prime blocks[32] directly on the southwest bank of the Murrumbidgee and adjoining Cavan to the east. These blocks were registered in 1873 and 1875 respectively in the name of Thatcher's infant son, William. Castle presumably financed the purchase of both blocks through Thatcher, intending to take legal ownership of them once the full purchase price had been paid. By using the name of Thatcher's infant son on the conditional purchase register, a loophole in the legislation that was later outlawed, Thatcher was preserving his right to select up to 320 acres for himself, as he began to do at Mountain Creek some years later, adding to the original landholding of forty acres that he had purchased from William Calvert.

The battle to keep Cavan together continued into the next generation after Castle's death in 1883 and was not always successful. For example, the two blocks on the river that were clearly acquired by Thatcher on Castle's behalf were ultimately transferred to a local resident, John Styles, in the early 1900s. Presumably, money was not available at the time to pay off these valuable river flats and the land could then be bought by someone else. Throughout New South Wales, this type of situation created a high degree of tension between long-standing squatters and new selectors under the Robertson Land Acts, though fortunately it seems that at Cavan, relations with the vast majority of conditional purchasers on or within its boundaries appear to have not only been cordial, but also close.

CASTLE, THE GOOD SAMARITAN

Castle had a reputation for having a kind and equable nature and his goodwill extended to all those around him. The impression from the Cavan diaries is that Thatcher especially was always welcome there, both receiving and lending a helping hand when required, even after he ceased his employment at Cavan and selected land for himself on the southern tip of the Cavan run. Similarly, this was true of the Archer family who settled on a block of fertile land just north of Cavan on the opposite side of the Murrumbidgee. John Archer is mentioned regularly in the Cavan diaries from the early 1880s, leasing additional land for himself at Cavan and frequently borrowing machinery for threshing and winnowing.

This apparently happy co-existence with new selectors is perhaps all the more remarkable given the widely differing backgrounds and social circumstances of the inhabitants of the Cavan district. Such divisions, coupled with the hardships of life on the land and the pressure to succeed, could easily have made for violence and discontent, but instead, the community grew and, in some quarters, prospered, from early settlement right through to the 1900s and beyond. Castle was an important centrepiece in this development, not only providing work for many people, but also campaigning vigorously for the establishment of a school and a post office. However, he was by no means the only person who represented the backbone of this diverse community; there were many others who contributed to the rich tapestry of the locality and gave it a notable place in Australian history.

CHAPTER EIGHT

THE CAVAN
COMMUNITY

CIRCA 1830 TO 1900

SOCIAL BACKGROUND OF THE INHABITANTS

THE COLOURFUL VARIETY OF PEOPLE that settled in the Cavan district from the early 1830s onwards is such that it could rival the cast of one of Shakespeare's plays or Chaucer's *The Canterbury Tales*. The social background of the inhabitants was by no means homogeneous; but contrary to popular imagination, it was not predominantly a convict community of labouring classes, though there would have been many such individuals working the land or in domestic roles for their more gentrified employers.

There is no official record of the early population of Cavan because the County of Cowley, in which Cavan is now situated, was only created in 1848. However, there is a very close alternative source, which is likely to give a fairly accurate picture of the Cavan community of this time. According to the 1836 Census of the County of Murray (across the river on Cavan's eastern boundary), there were 847 convicts out of a total population of 1,728. Of these inhabitants, 1,089 were Protestants, 630 were Roman Catholics, eight were Jewish and one person claimed to be a Pagan. In short, the early pioneers of Cavan and the wider district were not confined to any particular level of society, but were a diverse group both in terms of their socio-economic background and their religion.

Forming the nucleus of the Cavan community were the free settlers, educated men from largely middle-class or upper-middle-class backgrounds. Henry Manton fell squarely into this category as indeed did Edmund Lockyer, George Potter and, later, Joseph Castle. But there were also those such as Alexander and William Riley who represented the new breed of commercial men whose material success and upright characters

Above — *View from Bloomfield looking south over the Murrumbidgee and up the Mountain Creek flats*
Previous pages — *Constructing the new Taemas Bridge from the Cavan side, c. 1885*

earned them an added respectability and a degree of acceptance amongst the 'pure merinos' or pastoral elite.[1] This group, also known as the 'exclusives', was made up of men with social position and private means such as retired army officers, minor aristocracy and the younger sons of smart landowning families in England. It was the exclusives who fought the advance of the 'emancipists' (men who had either served their full sentence after being transported or been pardoned) and guarded jealously their positions of influence and power.[2]

The taint of convictry was significant and for many years this class sat below the free settlers. Even after emancipation, former convicts tended to be excluded from society and political positions, although some of them turned out to be the ablest men in the colony. Snobbery may have been rife in society, but the demand for convict labour in the bush was huge and convicts were readily employed as shepherds, shearers, ploughmen, blacksmiths and carpenters as well as domestic servants. Regulations under which convicts were assigned changed frequently, but about the time that Cavan was established, convicts were given sparse accommodation, clothing and rations (usually meat, flour, sugar, tea, tobacco and soap) and, depending on their employer, the recommended wage of £10 per annum. This was considerably less than that of a free man who, as a good stockman or shepherd, could earn £20 to £30 per annum plus rations. After 1823, paying a convict wages was not obligatory and was often used as an indulgence to encourage good behaviour and loyalty. Many who worked steadily rose to occupy all sorts of positions, including overseers and managers of large estates; if they had saved their wages, some even became landowners themselves.[3]

The colony in general was prospering 'on the sheep's back'. Although backgrounds in the Cavan community were vastly different, the majority of inhabitants in the vicinity of Cavan would have had in common a purpose to better their condition and to seize opportunities that they would probably never have had back in Britain. On the land, they knew that with a positive attitude and hard work the future lay under their feet, if only the often inhospitable landscape could be tamed to their advantage. The colony was now rewarding endeavour to a degree unknown back home, but to succeed on the frontier required drive and perseverance. Whether free settler or convict, man or woman, these migrants had stepped onto a threshold of hope, which, with luck and effort, ultimately promised everyone upon it independence and prosperity.

FREE SETTLERS

Hamilton Hume

Amongst the free settlers who originally established the Cavan area and its northern fringes was the explorer who is generally credited as having first discovered the district, Hamilton Hume. Unlike any of the free settlers who lived at Cavan itself, he was a native of New South Wales, a true 'Australian'. He was the eldest son of Andrew Hume, a former superintendent of convicts who arrived in Sydney as a free settler in 1790. After holding a series of other government posts (during which time doubt was cast upon his integrity on several occasions), and having married the daughter of another free settler, Andrew Hume became a farmer on a small block at Appin, about fifty miles south of Sydney. Hamilton was then about fifteen years old and it was from here that two years later this exceptional young man with a strong taste for exploration mounted a series of important expeditions

Hume, who moved to the Yass district in 1829, knew William Riley through their mutual sheep and racehorse breeding interests, the two men being of similar age with rural holdings in the same general area. Indeed, it is not hard to imagine that they were good friends or, at least, friendly rivals.

—

— Explorer Hamilton Hume

inland, including those to Bowral in the Southern Highlands, Lake Bathurst and the Goulburn Plains. His most famous journey was from Lake George to Port Phillip and back with William Hovell, between 1824 and 1825.

In recognition of his many discoveries, Hamilton Hume received a grant in 1828 from the government of 1,280 acres, and another adjacent block of 1,920 acres in 1830. These grants were on the Yass Plains, immediately south of Yass township, and were the beginnings of a pastoral enterprise that ultimately comprised about 18,000 acres, scattered throughout the district northeast of Cavan, around Good Hope in the County of Murray.[4] His main properties were known as Humewood, Marchmont and Euralie. Here, Hume established a Saxon merino sheep stud with stock bred from ewes purchased from the Rileys at Raby and rams chosen from the flock at Cavan.[5] Subsequently, Hume also carefully selected some Saxon merino rams from the flocks of Prince Lichnowsky and Baron Bartenstein of Silesia and imported them direct.[6] From the handful of letters that remain,[7] Hume, who moved to the Yass district in 1829, knew William Riley through their mutual sheep and racehorse breeding interests, the two men being of similar age with rural holdings in the same general area. Indeed, it is not hard to imagine that they were good friends or, at least, friendly rivals.

In 1839, Hume acquired the small freehold property Cooma, on the Yass River, described two years earlier by Thomas Walker in *A month in the bush of Australia*,[8] as having a 'very nice and commodious cottage, very well-furnished and with everything comfortable about it; the ground and garden nicely laid out, but as yet quite in their infancy'. Apparently, the impetus for this purchase was that Cooma lay on the very spot where Hume had camped during his overland journey to Port Phillip in 1824.[9]

Purchased from two other free settlers in the neighbourhood, the brothers Henry and Cornelius O'Brien, it was here that Hume settled with his wife until his death in 1873.

Henry and Cornelius O'Brien

Henry and Cornelius O'Brien were among the first wave of free settlers to occupy land northeast of Cavan shortly after the discovery of the area became known. Both from an Irish Catholic family, Henry settled in New South Wales in the late 1820s, eventually securing 15,000 acres of some of the best land on the Yass River and which adjoined the blocks acquired by Hamilton Hume on the Yass Plains.[10] He continued to occupy this estate, which he called Douro, for over thirty-five years, becoming the largest stockholder in the district and a prominent member of the community.

Henry was a member of the Yass District Council in 1844 and in 1860 was elected unopposed to represent Yass Plains in the new Legislative Assembly.[11] Also blessed with entrepreneurial skills, Henry was the first grazier to devise the scheme of boiling down livestock for tallow, a practice he developed during the financial depression of 1843 when the value of livestock was practically nil.[12] This practice became a godsend to pastoralists in times of drought and hardship. The boiling down plant stood on the banks of the Yass River at Douro until it was washed away in a flood in 1860. Like Alexander Riley, Henry was keen to improve the quality of his wool and, according to William Riley, crossed his sheep with pure Riley Saxon rams so that some of the third cross might be mistaken as the genuine breed.[13] Allegedly, he also imported sheep from the Chatillon Stud in France at considerable expense to enhance his flocks which, in William

Riley's opinion, became superior to those of his neighbour, Frederick Manton at Mon Reduit.

Douro had a superb view of the distant mountains, but according to William Riley, the house was 'not the most comfortable' in the colony, being built of logs and very draughty. The land on the plains, however, was covetable and included some of the pastures that William Dutton originally applied for on behalf of Alexander Riley, but which at the time, fell within a church reserve and were thus refused. The property was described presciently by George Bennett[14] a year or so after Dutton's failed application as generally 'excellent quality' and being well watered, likely to 'prove the richest and most valuable of almost any of the present known portions' of New South Wales. In poetic mood, Bennett waxed lyrically about the landscape: 'The plains were animated by herds of cattle, flocks of sheep, and troops of horses, grazing, reposing or exercising; the whole combination of this beautiful scenery excited the most pleasing sensations, which were heightened by its English character, and cause the settler to reflect less on the remoteness of home.'

As was common with many settlers, Henry's brother, Cornelius, joined him, arriving in the Yass district in 1833 and taking up land adjoining Douro. Together, the brothers also ran a large cattle station, Coppabella, on the Murrumbidgee, and further south at Jugiong, an area known at the time as being 'on the extreme edge of civilisation' and haunted by 'wild and troublesome' Aborigines.[15] In addition, Cornelius owned Hardwicke at some stage, possibly purchasing this from William Dutton with whom he was likely to have been acquainted. Both O'Brien brothers contributed much to the development of the Yass district and, like the Rileys, were pioneers in the opening up of new country and the general improvement of wool quality in the pastoral industry.

Rees Jones

Another significant figure in the Cavan area was Rees Jones. A free settler, he arrived in Yass about 1849 and, initially opening a general store, later established a soap-making business and a flour mill in the town.[16] Customers often paid him in wheat, which he stored in the shearing shed belonging to Henry O'Brien.[17] Jones had previously been successful in trade in Sydney and was the brother of David Jones, who founded the department store of the same name, which still exists today. Also acquiring land and grazing licences in the Hunter and Queanbeyan regions,[18] Rees Jones was declared insolvent in 1842, a likely casualty of the prevailing economic depression.[19] Nonetheless, he seems to have recovered his finances whilst in Yass and, in 1854, purchased the 16,000-acre property of Taemas, between Cavan and Narrangullen. He lived here for a short time before returning to Yass, leaving his wife and family on the farm and visiting them only at weekends.[20]

By this stage, Jones was a prominent member of the Yass community, acting as Vice President of the Yass Pastoral and Agricultural Society for many years and becoming both a freemason and mayor.[21] He was also on the bridge committee for the first bridge over the Murrumbidgee at Taemas and would have known his next-door neighbour at Cavan, James Calvert, who championed the idea. Aside from his numerous activities in town, Jones expanded his pastoral interests at Taemas, acquiring the adjoining property of Narrangullen from Alexander Raby Riley on 31 July 1862 for £2,000.[22]

Jones's wife, Ann, was more homely than her husband and wrote an account of her marriage,[23] describing in detail the lonely and miserable existence that she had on Taemas whilst her husband availed himself of masculine sociability in Yass.

His myriad of interests kept him from home a great deal, which made his wife unbearably sad; he 'gossiped in the banks, lunched at the hotels and every now and then at the Free Masonry meetings [was] out til sometimes 3 in the morning'. According to one local banker, he was such a clever and knowledgeable man that 'Yass could not get on without him'. With such extraordinary popularity amongst the community, a special dinner was given for him in 1861 by the townsfolk to mark his retirement from business. Nonetheless, Jones must have been an insensitive and selfish husband as, in spite of his wife's pleading, he refused to buy her a cottage in town so that she could at least live with him there during the week.

Although a success in Yass, Jones had a mixed experience at Taemas. He appears to have been in some kind of trading partnership with a man named Gorman, who initially had the full management of the place before the Jones family moved there and was entitled to 'have every other calf branded in his name'. Gorman resented the family residing there as well as Jones's weekend visits once Jones moved back into town. According to Ann's memoirs, Gorman was 'an awful liar' and made himself 'very obnoxious' so that upon her husband giving him notice, he sued Jones and 'there was no end to the annoyance and expense'. The case dragged on for some years and, to settle the parties' differences, the court ordered Taemas to be sold. However, with some money that Ann had put aside, the Joneses managed to retain the property after it was advertised for sale in 1858;[24] and 'Gorman who was drunk was found dead on the road with both legs broken some few years later'.

Ann's memoirs come to an abrupt end about this time, although it was not until 1886 that her husband died at Taemas. Perhaps the intervening years were too painful to record or maybe she adjusted to her circumstances and made the best of

— *View towards Bloomfield showing remains of original Taemas Bridge in the foreground*

them. With the Castle family close by at Cavan, it is just possible
that she at last had some company to call upon besides that of
her two sons, Hamlyn and Joe, who appear to have been the only
ones out of her five surviving children to have remained at home.

EMANCIPISTS

Samuel Terry

Of equal fame to the Rileys at Cavan, but from a very different background, was their 'neighbour' and absentee landlord, Samuel Terry, whose property, Kenilworth, also lay to the north of Cavan on the Yass Plains, on the eastern border of Henry O'Brien's property of Douro. Terry had been a labourer in Manchester, England, and was a former convict, transported to New South Wales in 1801 for the theft of 400 pairs of stockings. In this respect, he bore more resemblance to James Fitzpatrick, the earliest possible occupier of the Cavan run, than to its later owners. Once emancipated, however, he prospered rapidly through innkeeping and then by speculation in both urban and pastoral properties. Becoming extremely rich and earning the epithet 'The Botany Bay Rothschild', he was regarded as the leading emancipist of his day, having spectacularly reversed his earlier adversity.[25] An agitator for political and social reform, including the establishment of a house of assembly and trial by jury, Terry was regularly appointed to be treasurer, or director, or president of political, social and business organisations, due to his ability to organise and fundraise, as well as to his sober and industrious personality. Terry was also a philanthropist, supporting a myriad of causes, including the founding of Sydney Free Grammar School, the forerunner of today's Sydney Grammar.[26]

In spite of his good works and enormous wealth, Terry's status as an emancipist largely excluded him from the social life of the colony, which in the early days of settlement was dominated by free settlers and British officials. The persistent rumours about Terry's lack of integrity and ruthlessness in business would also not have helped his path into polite society. Although he had become a member of the merchant class to which Alexander Riley belonged, it is very unlikely that the two men would ever have socialised together. In society, the line between the free and the freed was clearly defined.[27] However, in the world of business, it was barely visible. The two men certainly knew each other and met from time to time through commercial circles in Sydney; Terry had been exporting goods from New South Wales in his own schooner from as early as 1817 and Alexander Riley gave evidence in the Bigge Report of 1823, into the state of agriculture and trade in the colony, that Terry was an importer of goods from India and Europe.[28]

By 1828, Terry had ventured into rural property, becoming an extremely successful pastoralist and one of the greatest landholders in the colony. He owned more than 20,000 acres across properties on the Nepean River and at Liverpool, Bathurst and Yass, as well as innumerable sheep and 3,700 head of cattle. In 1832, he augmented his stock further by the purchase of all cattle in Appin and Shoalhaven belonging to another notable landowner, the late D'Arcy Wentworth. William Riley mentions Terry in his journal, which he kept during his travels to the Yass Plains in 1830,[29] claiming that at Kenilworth Terry had 800 head of cattle and a 'scurrilous' stockman, George Davis, a former convict, who sold rum on the black market for ten shillings a bottle, keeping what was popularly known as a 'sly grog shop'. Perhaps Davis had learnt a few things about business from his absentee employer who was happy to turn a blind eye to his stockman's initiatives.

George Davis

In 1838, Davis purchased 505 acres of exceptional alluvial flats on the Murrumbidgee for £707.[30] Having already acquired land at nearby Gounyan, this latter purchase suggests that Davis had prospered during his employment with Terry and been able to build up some capital. He had put behind him his earlier sentence to seven years' transportation for assisting the escape of French prisoners from British occupying forces and, like Terry before him, had turned adversity to advantage. The beautiful property on the river bank was almost directly opposite Cavan and just to the south of land owned by Terry's son-in-law, John Hughes. Davis named his acquisition Bloomfield, a reference perhaps to the wealth of flora growing in the luxuriant pastures that fronted the river for almost a mile along the whole of the western side. Today, Bloomfield is part of modern Cavan and is still the idyllic location that attracted Davis to the site.

Rather like his former employer, Davis became a wealthy man. He converted his 'sly grog shop' on his property at Gounyan near Yass to an inn, The Sawyer's Arms, and purchased various other pastoral properties in the district, including land at Mundoonan and Sutton and a cattle station, Greenhills, further south at Adelong. About 1853, Davis gave Bloomfield to his youngest son, Thomas, and his wife, Elizabeth, who was of a literary bent and wrote a little uncharitably in her memoirs that 'it was the mistake of my life to marry an uneducated man'.[31] Apparently, Thomas was illiterate, as were many others of his generation in the area, there being no government-funded schools in the district at this time – not even in Yass, where the first public school was only established in 1872.

Thomas and Elizabeth built a four-roomed house of fieldstone, which still stands today but now with many recent

— *Bloomfield homestead today*

additions. A fire in 1854 destroyed the kitchen, dairy and storerooms, but the house was saved. Nonetheless, the young couple decided to move elsewhere and sold Bloomfield in 1860 to a doctor, Allan Campbell,[32] who later also purchased the neighbouring property, Warroo, belonging to Alexander Raby. Campbell installed Arthur Webster as Bloomfield's station superintendent. Webster lived at the property for twenty-seven years, had a large family and was active in community affairs, vigorously supporting the application for the first Cavan school.

Samuel Taylor

Samuel Taylor was another early settler in the area, also of convict background. He was a labourer, arriving in Sydney from Kent in 1820 on the *Coramandel*, having been sentenced to seven years' transportation for an unnamed crime. He was given his freedom on 27 July 1826. Clearly, he was rather a character, known in the Yass district for his hard riding and drinking. According to an article in the *Yass Courier*, 'Sam had two distinct traits of character, one strong, the other (mentally) weak. He loved a good strong horse. He also loved strong rum, which transformed a shrewd horse dealer, when sober, to a tantalising menace to the peace and safety of civilians when he came to town for a "blow-out".'[33]

Charged several times for reckless riding in the Yass township,[34] Taylor was a talented horseman and there were few who cared to follow him down the thickly timbered slopes around Mountain Creek where he chased and mustered brumbies, displaying the finest bush-riding skills. On one occasion, he drove a mob of horses all the way to Adelaide without a single loss in order to obtain better prices than he could locally. When asked how he had managed this, he replied, 'No pubs within a day's ride of the track'!

In spite of his rough ways, he was a successful pastoralist and owned Taemas[35] for a time before he sold it to Rees Jones, allegedly for £2,500, a considerable sum in those days. Like Samuel Terry, Taylor was an emancipist, and the two men appear to have been entirely different in their aspirations: Terry became a noted public figure in contrast to the bushman and vagabond image that Taylor cultivated. No doubt, there were many amongst the exclusives who felt both Terry and Taylor were one of a kind, whatever the different lives and outward appearances of the two men might have been.

SELECTORS UNDER THE ROBERTSON LAND ACTS

In 1861, land legislation in New South Wales was completely reformed by the introduction of the Robertson Land Acts. Whilst this new legislation sought to protect existing pastoral interests at the same time as make way for new settlers, it ultimately divided the rural population and encouraged a class war for possession of land. Each group resorted to all sorts of chicanery from 'dummying' to bribery of officials in the lands department and to physical violence. Fortunately, corruption and brutality does not seem to have been the rule around Cavan. The many selectors that came to the district post the introduction of the new land legislation included John Thatcher, John Archer and John Styles, all of whom were employed at one time or another at Cavan. There did not appear to be any animosity between them when they became selectors; the Thatchers, Archers and also the Grace family clustered around the southern end of the Cavan run along Mountain Creek, whereas Styles established himself to the east on the Murrumbidgee next to the original Cavan grant. There is no doubt, however, that entries in the Cavan diaries in the 1880s and beyond show a preoccupation by the owners of Cavan with securing as many conditional purchases for themselves as possible out of the original Cavan run to protect their landholding from the influx of new occupiers.

Notwithstanding the good intentions behind the Robertson Land Acts, free selection by conditional purchase was a mixed success. The limitation on the amount of land that a selector could hold at any one time (originally 320 acres and increased in 1875 to 640 acres) meant that creating a viable farm or grazing business was difficult because the amount of land was too small an area to be productive in times of drought and, in any event, had restricted carrying capacity due to the particular fragility of the wild grasses. Tragically, many selectors in New South Wales lived in appalling conditions, in rough bark huts with little food and no sanitation, working night and day to keep themselves and their families alive. The vagaries of climate and prices, ignorance of farming in New South Wales compared with the British Isles, and inadequate or non-existent transport to market, all contributed to their penury. When unable to meet their financial commitments, their creditors foreclosed and many selectors were forced to work in the district for wages on larger, established properties such as Cavan or to move to the towns. Their life was one of hardship, not made any easier by the lack of infrastructure in the wilderness. This was slow to evolve and, when it did appear, it was not always the success that had been anticipated.

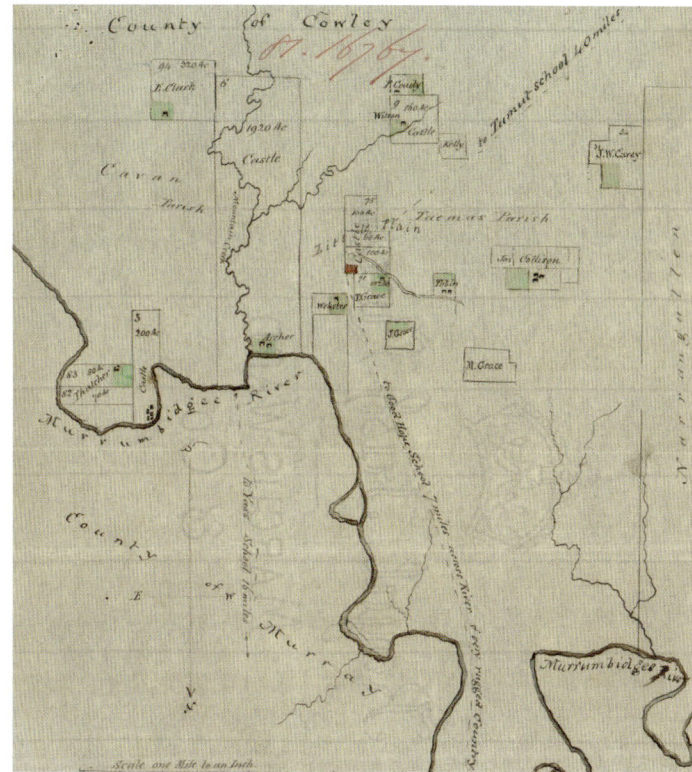

— *Plan of school site drawn by Joseph Castle*

CAVAN SCHOOL

It was a very long time before government-sponsored education was made available to the scattered Cavan community, although there were two or three private church schools in Yass from the late 1840s.[36] Not until after the new land legislation in the 1860s increased the population by its promotion of closer settlement, could any justification be made by the Department of Education for a school in the locality. Unsurprisingly, the champion of this initiative was Joseph Castle, former schoolmaster and now the owner of Cavan, one of the most prominent graziers of the district. He was supported in his efforts at various times by several locals, including the grazier Charles Hall of Yeumberra; the free selectors Edward Clark and John O'Donnell; Castle's

nephew, Andrew Roche; and the superintendent at Bloomfield, Arthur Webster.

By the late 1870s, the population of Cavan had grown to one hundred, including thirty children of school age living within a three-mile radius.[37] According to a letter dated 23 June 1881 from Castle to the Department of Education,[38] most of the populace were either 'struggling' free selectors or labouring men, employed by the graziers and earning a meagre 15 to 20 shillings per week. Spurred on by the considerable poverty in the area and the desire to improve the prospects for the younger

— Joseph Castle's application for a public school at Cavan

members of the community, Castle lobbied the authorities for ten years, from the early 1870s, until they finally agreed to open a provisional school at Cavan in February 1882.

The main reason that the application for a school was declined for so long was that most parents were obliged to keep their children at home working, tending sheep and cattle from a very early age, but some were also reportedly 'apathetic and indifferent'[39] as to whether or not their offspring received an education at all. It was therefore doubtful that attendance at the school, were one to be established, would be high or regular. A compromise of a half-time school at Cavan was initially offered by the District Inspector with the option of attending the provisional school at Good Hope, about ten miles north across the Murrumbidgee, which had opened in 1872. This suggestion does not appear to have been acceptable to the local inhabitants and, in any event, the provisional school at Good Hope closed in 1878 so that the only other schools in the district were at Spring Creek, established in 1873 and sixteen miles away, or at Yass, which was even further.

The turning point seems to have been when Castle offered up some of his own land for the school on the western side of Cavan 'on a dry limestone plain with a spring of water near it'[40] and a slab hut which was duly completed in time for the school to open in February 1882. Four of the maximum twenty-six children who attended were John Thatcher's, the long-standing employee at Cavan; and the first teacher was William Bullen, who was much respected and stayed four years.[41] During this time, he married one of the daughters of Arthur Webster, the superintendent at Bloomfield. The residents of the area were unable to afford the provision of a house for the schoolmaster, but Bullen was nonetheless, 'cheerfully boarded and lodged by the parents of the children'.[42] When Castle visited the school in the winter of 1882, he noted that daily attendance had averaged twenty-two and was on the increase.

Although the school itself proved a success, the slab hut was bitterly cold in winter so that Castle began a campaign for the government to pay for its improvement or replacement. After some argument, not only was a new school erected a short distance away in 1885 on land[43] provided by the free selector Thomas Grace, but it was upgraded to a public school because of the increased number of pupils. Here, the school stood close to the junction of the Wee Jasper and Mountain Creek roads. Castle had died two years earlier and so did not know how his persistence had paid off. The school closed in December 1895, probably due to a lack of students (the original cohort having grown up) and to the rural depression prevailing at the time, which would have meant that families either moved away or kept any children at home for chores.

THE CAVAN POST OFFICE

Before the school came the post office, a lifeline to the community that otherwise had no means of communication with the world beyond their district. Once again, the main protagonist for the establishment of a post office was Joseph Castle.[44] Upon the closure in 1876 of the post office eight miles away at Taemas, Castle petitioned for a post office in the centre of the Cavan run, recommending the free selector John Carey of Gum Flat as postmaster.[45] Carey was said to have had a room in one of two bark huts available for the purpose, and his son, William, could fetch and carry the post once a week on horseback between Cavan and Yass via Brassill's Inn at Warroo, a journey of about four hours.

The populace were either 'struggling' free selectors or labouring men, employed by the graziers and earning a meagre 15 to 20 shillings per week. Spurred on by the considerable poverty in the area and the desire to improve the prospects for the younger members of the community, Castle lobbied the authorities for ten years, from the early 1870s, until they finally agreed to open a provisional school at Cavan in February 1882.

—

After some dispute as to whether a post office at Cavan was necessary, the post office was opened on 1 March 1877 with John Carey in charge. The Yass and Taemas postmasters had claimed that only twenty people would use such a facility, whereas Castle had persuaded the authorities in Sydney that more than a hundred people would benefit. He managed to collect about sixty signatures from the heads of local families to prove this, saying that he could have doubled this number if shearing had not been in progress and there had been a single idle man available to obtain more names.[46] Amongst the signatories were Castle's nephew, Andrew Roche, and Arthur Webster of Bloomfield.

The post office seems to have run smoothly given the lack of complaints on file, and the only real issue arose in 1885 when John Carey resigned as postmaster on the grounds that he was leaving the district. The post office was then moved about two

Scenes of Bloomfield,
from c. 1890–1910

miles away to the residence of a local farmer, John Collison, who was paid £10 per annum for a weekly mail service. Collison resigned in December 1901, whereupon the post office closed for lack of a suitable successor. On 15 December 1913, it reopened as a receiving office, operated by Thomas Grace on his property at Little Plain, eight miles from the original post office in the centre of the Cavan run. For how long this facility remained is unclear, as the relevant documents are missing from the official file, but with the advent of the motor car and road transport, it is likely that it did not survive for very long as the full post office in Yass would have become easily accessible.

YASS TOWNSHIP

By the time of the first Cavan post office and school, Yass had become a thriving frontier town. About eighteen miles from Cavan, it was still too far for a round trip in one day but was nevertheless an important centre for supplies, banking, rail transport (at Yass Junction) and communications. It also had a number of churches which were lacking in the immediate Cavan area, attracting settlers in the outlying districts for landmark occasions such as baptisms, marriages and funerals. Lying on the banks of the Yass River, a tributary of the Murrumbidgee, Yass was on the main route from Sydney to Melbourne, and its prominence in the growing of fine wool and also its wheat industry ensured the town's transformation from a tiny village to a significant commercial and social hub.

Early activity in Yass, which was originally founded around 1830 but only officially gazetted on 4 March 1837, was very limited.[47] A government poundkeeper for lost stock was established in 1833 and a courthouse was set up in 1834. As Yass

By the time of the first Cavan post office and school, Yass had become a thriving frontier town. About eighteen miles from Cavan, it was still too far for a round trip in one day but was nevertheless an important centre for supplies, banking, rail and communications.

—

became the administrative centre for the County of King, Henry Bingham was located there. Bingham was the Commissioner for Crown Lands for the Murrumbidgee district who was at the centre of the dispute over the loss of the Cavan run to Horatio Beckham in the early 1840s. The first general store in Yass began trading in 1838 and two inns were opened: The Rose in 1837 and The White Hart in 1839. The first mail coach arrived in Yass in 1835 over a track from Goulburn linking the town to Sydney, but it was not until 1865 that the mail run went all the way south to Melbourne. The telegraph, however, reached Yass in 1858, a pivotal moment in its history.

Religion came to Yass from 1833 onwards with visits by ministers of various denominations, including the missionary journey of the Catholic priest Reverend John Joseph Therry, who recorded in his diary calling on Henry and Cornelius O'Brien and the Mantons.[48] The O'Briens were the only Catholics of substantial wealth and social standing in the district at the time; and the Mantons, although Protestant, may have had Catholics amongst their convict assigned servants. The first Catholic church in Yass, St Augustine's, was built about 1840, although its foundation stone was blessed and laid already in 1838; the earliest Protestant church, St Clement's, was not properly established until 1850, Anglican services having previously been conducted in the courthouse and the subscription library.

Gradually, the population of Yass grew from just 37 people in 1837 to 141 in 1841 and to 270 in 1848. By this time, the town had several flour mills for processing wheat. A gaol was established in 1861, a bank in 1864, a mechanics institute and reading room in 1869, and a public school in 1872. The railway to Yass came in 1876 and, although the station was a few miles out at Yass Junction, it further connected the town to Sydney and, later, to Melbourne.

As time went on, Yass appears to have played an increasingly important part in Cavan life, its occupiers spending more time there both for social and business reasons. This started to become evident in the generation that followed Joseph Castle, but Cavan still remained mostly self-sufficient until the turn of the century. Nonetheless, it was a phenomenon occurring all over New South Wales towards the end of the nineteenth century that, contrary to what had been imagined by the politicians, many people clung to provincial towns rather than lead a life of relative freedom in the bush. As Henry Lawson lamented in his poem 'The Roaring Days', a tribute to the golden era of pioneering in the bush, the frontier was changing as miles of rail tracks were laid down to connect country and town. He wrote vividly and with a high degree of realism, 'The flaunting flag of progress is in the west unfurled, the mighty bush with iron rails, is tethered to the world.'

Above top — *New Taemas Bridge today built post the flood of 1925*
Above — *View of Cavan with uprights of old Taemas Bridge just visible*

Above — *Near Taemas Bridge with cattle, c. 1890*
Opposite — *Opening ceremony of the first Taemas Bridge, 1888*

CHAPTER NINE

THE ROCHES

THE DEATH OF JOSEPH CASTLE

CASTLE, THE ABLE SCHOOLMASTER turned pastoralist, died at Cavan on 22 May 1883 at the age of seventy-two after an illness of six months. He was by then a wealthy man and highly esteemed by all who knew him. His funeral was numerously attended and his remains were interred at Cavan.[1] He left behind his formidable wife, Wilhelmina, and his as yet unmarried daughter, Eliza. In his will, Castle bequeathed everything, including Cavan and some small landholdings in Wollongong, to Wilhelmina and Eliza in equal shares as tenants in common, envisaging perhaps that on her death Wilhelmina might choose to leave her share to Eliza's husband if she were married. Castle's executor was his brother-in-law, Frederick Roche, with whom he had for a short time been in partnership in Queensland, and the estate was valued at the impressive sum of £14,112.[2]

ELIZA CASTLE

Although Castle left Cavan to his wife and daughter in equal shares, it appears from the Cavan diaries[3] that it was Eliza who took charge of the management of the property and that her mother was not involved, except perhaps in the day-to-day running of the homestead. In spite of her dominant role at Cavan, unusual for a woman in those days, Eliza remains an enigmatic figure. Few documents concerning her survive, beyond

Left top — *William Frederick Roche (Fred), c. 1885*
Left — *Canoeing on the river at Cavan, c. 1915*
Previous pages — *Laddie Castle-Roche (far right) at Cavan with two unknown gentlemen, c. 1915*

her birth and death certificates, a small number of letters from her husband, a larger bundle of correspondence from family and friends, two photographs and a very short obituary. Nor have any stories about Eliza been handed down the generations; the only personal possession of hers known to exist is a small album or scrapbook containing a handful of photographs of Cavan and some favourite poems and cards.[4] Although Eliza was a keen correspondent, judging by the letters she received from relatives and friends in Ireland, England and from all over Australia, not a single letter in her own hand and revealing information about herself has found its way into any family papers or archives. With so little material, only the bare facts of her life can provide any insight into her true personality, which was evidently an intimidating one, not dissimilar to her mother's.

It is clear from the very affectionate tone of her father's few remaining letters that Eliza was his pride and joy and that the two were very close. This was true from the time that Eliza was a little girl to when she became a young woman. In a letter dated 2 May 1870,[5] Castle wrote to Eliza, who was staying with friends in the Southern Highlands: 'My wonderful girl, I am this moment in receipt of your note of Friday in which you speak of coming home this week. I am half afraid you will never get permission to go away again after this long absence – unless indeed you take it into your head to run away altogether, an event which I dare not anticipate.' As with all his letters, Castle gives Eliza a vivid account of family friends he has seen or heard from and of activities at Cavan or in Sydney that might be of interest to her. Always signing himself as 'Your loving Papa', there can be little doubt that father and daughter were devoted to each another.

As his presumptive heir and the future chatelaine of Cavan, it is fair to assume that Castle took care to educate Eliza

As his presumptive heir and the future chatelaine of Cavan, it is fair to assume that Castle took care to educate Eliza well beyond the normal expectations for women at the time, especially given his own background in education and his achievements in that field.

—

well beyond the normal expectations for women at the time, especially given his own background in education and his achievements in that field. There was an extensive library at Cavan which, according to Castle's great, great granddaughter, Susie Castle-Roche, contained a large number of books with Eliza's name written on the inside and which were bequeathed to the National Library sometime in the mid-1960s. Some of these books were rare and included histories of the ancient world written in French. What Eliza's reading tastes were is unknown, but her father implied in one of his letters that she was able to read ancient Greek. It is unclear where Eliza received most of her education – she grew up in Sydney and the family did not move to Cavan until she was twenty-one, but several of her father's letters appear to have been written to her whilst she was away at boarding school, though no name or address for the establishment is given on the correspondence and the envelopes have been lost or destroyed.

With this upbringing and the prospect that Cavan would someday be hers, it is probable that Eliza was unusually independent, a fact that is supported by her marrying late and by her sharing the management of Cavan with her future husband for four years before they were married and for some months afterwards. Whether Eliza was interested in women's rights and the growing suffragette movement of the period is unknown, but her father was a close friend of Walter Hurndale Windeyer, an uncle of William Charles Windeyer,[6] whose wife, Mary, was a noted suffragist, notorious for her endeavours to advance women's opportunities and political rights. She was also of the same generation as Eliza with whom she may have been acquainted through family connections.

SUITORS

It seems that Eliza took her time to choose a husband, not marrying until the age of thirty-eight. Although she had been plucked from her urban life in Sydney at twenty-one, just at the time she might have expected to be meeting her future spouse, being transplanted to Cavan would not necessarily have limited her marriage prospects. By 1870, when the family came to Cavan, Yass was a thriving town with a population of over 1,200 people and the whole district was becoming increasingly settled. Even the mainline railway reached the outskirts of Yass in 1876, so that it was possible to travel to Sydney in less than six hours. Moreover, there were regular race meetings at Yass as well as other social events such as musical and theatrical evenings that would have provided opportunities for Eliza to meet people of similar background.

However, it was her Roche cousins that seem to have been her main suitors, whether by her father's design or natural inclination. First, there was Andrew Kerr Roche, the second son of Frederick Roche (Wilhelmina Castle's brother) and his wife, Helen. Early in the 1870s, Andrew seems to have

worked at Cavan for a time before returning home to Dalby in Queensland, where his father had become the town's first mayor and was now a successful storekeeper. On 15 January 1875, Andrew wrote affectionately but tentatively to Eliza saying, 'I would like to go back to Cavan do you think that Uncle would take me back again if so I would try my best to please him, but if he does not feel inclined would you write by return of post and tell me as there is a new bank opened in Dalby which I would try and get into as a junior clerk.'[7]

Whether Andrew's main interest was to gain work experience or to court his cousin is not clear, but he seems to have received an encouraging response from Eliza. He wrote eagerly in his next letter of 2 February 1875, 'My dear Eliza, I received your letter last night and was very pleased to get it. Believe me, I shall be very happy to accept Uncle's kind offer. You will not I hope have reason to regret inviting me back at any rate I will do my best to please you all.'[8] Andrew returned to Cavan in April that year and, according to Castle's work diaries, was often seen riding out with his young cousin, Eliza.

However, although Andrew was employed at Cavan for several years, he did not win Eliza's hand if that was his aim. Instead, in 1881, he married Harriet Hall, one of the five daughters of Charles Hall who lived with his brother, Henry, at nearby Yeumberra, east of Cavan in the neighbouring parish. There are several mentions in the Cavan diaries of visits by the Castles to Yeumberra so that the two families must have been on familiar terms. (Charles Hall even appears to have named his only son, Charles Castle Hall, presumably after his friend and neighbour.) In 1879, a few years before his marriage, Andrew purchased 320 acres between Cavan's western boundary and Spring Creek Station (part of the Cavan estate).[9] Given its location on the Cavan run, this purchase was probably on behalf

By 1870, when the family came to Cavan, Yass was a thriving town with a population of over 1,200. Even the mainline railway reached the outskirts of Yass in 1876, so that it was possible to travel to Sydney in less than six hours. Moreover, there were regular race meetings at Yass as well as other social events such as musical and theatrical evenings.

—

of his uncle, acting for him as a 'dummy', and later transferred into Castle's name when the full purchase price had been paid. Andrew's own property, Ravenswood, which is now part of Cavan and where he lived with his wife and children, was just over the river to the east of the original Cavan grant. It had probably been part of Yeumberra[10] and may have been given by Charles Hall to Andrew and Harriet upon their marriage.

The spanner in the works as far as any romance developing between Andrew and Eliza was concerned may have been the visit to Cavan by Andrew's eldest brother, William Frederick Roche (known as 'Fred'), in October 1876, a year or so after Andrew had begun working there. Again, several of Castle's diary entries mention Eliza riding out with Fred during his month-long stay. Where he went immediately after his visit to Cavan is not known, but by late 1880 Fred was droving cattle around the remote region of Lake Nash in the Northern Territory, recovering from a debilitating bout of scurvy and

writing regularly to Eliza. His letters are serious and dull but kind, mainly answering Eliza's anxious enquiries about his health, telling her that he is 'quite well again though stiff and slow', and apologising for the lack of any news to interest her beyond his simple yet extremely taxing life outback.[11]

But Eliza was not the only one concerned for Fred's well-being. Her father seems to have been very fond of the young stockman, writing to Eliza on her visit to her Roche relations in Dalby in August 1881, not to allow her uncle to complain that Fred had sold his cattle in the north as it was 'more important that he returns and is restored to health'.[12] Castle also expressed the hope to Eliza that Fred would arrive in Dalby before her departure for Sydney, where father and daughter were to meet, saying, 'don't shorten your trip on my account'. Was this a father's ruse to bring these young people together?

By October 1881, Castle was not so backwards in coming forwards. He offered Fred the management of Cavan, presumably in the knowledge that by doing so, there would be a strong possibility of Fred becoming his son-in-law. According to a letter of 7 October 1881 to Eliza from George Roche,[13] the third Roche brother, Fred did not accept the job straight away, his object being 'to say nothing until he sees you all and then surprise you'. In the only contemporary description that remains of Fred other than his obituary, George describes him as a workaholic, a fact that is evident from Fred's own letters written from Lake Nash, warning Eliza 'you will have to watch him or he will be … knocking himself up'. Apparently, on his way from Dalby to Cavan where he arrived on 18 October 1881, Fred sensibly intended to treat himself to a Turkish bath!

It was a long courtship. Fred was five years younger than Eliza and of a very different thread. Whilst they were first cousins, he was plainly a man of the land – practical, down to earth and with simple tastes; Eliza, on the other hand, was educated and cultured, used to the intellectual company of her father.

—

MARRIAGE TO WILLIAM FREDERICK ROCHE

In spite of what appears to have been parental approval and mutual affection between the young couple, it was a long courtship. Fred was five years younger than Eliza and of a very different thread. Whilst they were first cousins, he was plainly a man of the land – practical, down to earth and with simple tastes; Eliza, on the other hand, was educated and cultured, used to the intellectual company of her father and also, up to a point, of her mother, an intelligent woman and former governess. Eliza was also potentially much richer than her hard-working, less worldly cousin, and it is possible that both parties were a little afraid that the compromise all round would be too great.

If these considerations were obstacles, they were finally overcome when the couple married on 19 March 1887, after Castle's death and five and a half years after Fred came to live at Cavan as manager. Interestingly, there is no evidence at

this time that, in line with convention, the couple adopted a double-barrelled surname to preserve the Castle name. Fred had been baptised William Frederick Castle Roche and signed his marriage certificate as 'F.W.C. Roche'. However, by the time of his death, the hyphenated surname of Castle-Roche seems to have been in use according to the various death notices and obituaries, but it is unclear when this first began.

Importantly, Eliza asserted her independence right up until her marriage, keeping the Cavan work diaries herself, with occasional entries made by Fred. A certain reluctance to relinquish the reins of Cavan is discernible, especially since it was not until early 1888, about nine months after their marriage, that Fred seems to have been permitted to make most of the diary entries himself. Possibly it was Eliza's pregnancy at the time that caused her to loosen her grip on Cavan. The couple's only child, William Frederic Fletcher Castle (known by his parents as both 'Lad' and 'Laddie'), was born on 17 August 1888 in Goulburn. His surname was registered as plain 'Roche'. Several affectionate letters from Fred to Eliza at this time survive,[14] indicating that the couple were close and that Fred especially was excited by the birth of the new 'squatter' and 'our little poddy'. Nevertheless, Fred had a rather gauche way of expressing himself at the news of the new baby, saying in a letter of 18 August 1888 that he was relieved it 'was not born in a curious way'. In any event, the arrival of a son and heir galvanised Fred into action and he wrote to Eliza in the following weeks that he had been decorating a room for the baby with calico, arranged godparents for his son and bought a new cow with a young calf to provide him with milk.

On her return to Cavan after the birth, there are references in the Cavan diaries to Fred taking 'Sweetheart and the baby' out for regular drives. Even if Eliza was distracted by motherhood, she appears to have maintained a strong interest in the Cavan enterprise throughout her marriage. Letters from Fred to Eliza in later years confirm that she retained this involvement in Cavan until her death; it was her personal tragedy that, due to ill health, she was forced in the last years of her life to be away a great deal from the place that, beyond her family, had been her main interest.

KEEPING CAVAN TOGETHER

Reading the Cavan diaries, Fred appears to have lived and breathed the place, working every day of the year, except Christmas Day and the occasional Sunday. He was rarely off the property, except for when conducting business in Yass, and appears not to have taken a holiday in the first ten years. Under his auspices, Cavan flourished sufficiently to withstand the impact of the Robertson Land Acts, the great depression of the 1890s and several periods of drought. This success was, however, not without some risk-taking and juggling of stock numbers. In 1888, Fred noted in the Cavan diaries that there were 20 horses, 28 shorthorn cattle and 8,000 sheep on a total acreage of 33,587, comprising 2,707 acres of freehold land, 3,570 acres held under conditional purchase and 27,310 acres under Crown lease (being part of the original Cavan run). Pursuant to the *Crown Lands Act 1884*, the Cavan run had by this time been roughly divided into two equal parts: an area of resumed land that was available for selection under the conditional purchase scheme and another area that remained a pastoral lease. According to an official register of pastoral possessions in New South Wales of 1889, an estimated area of 19,200 acres south and southwest of Narrangullen became resumed land whereas about 27,310

— *Kangaroo hunt, with Laddie at centre, 1908*

Under [Fred Roche's] auspices, Cavan flourished sufficiently to withstand the impact of the Robertson Land Acts, the great depression of the 1890s and several periods of drought. This success was, however, not without some risk-taking and juggling of stock numbers.

—

— Fence building at Cavan, c. 1908

Two valuable blocks on the river flats were bought by John Styles in the early 1900s. Since these were extremely covetable, the most plausible explanation for their 'loss' to Styles is that the Roches could simply not afford the conditional purchase, given all the other calls on their capital.

—

acres east, south and southwest of the original Cavan grant was designated leasehold.

By 1893, although sheep numbers had increased to about 9,000, there had been a further dramatic decrease in the Cavan landholding to 10,394 acres, almost the whole of the remaining leases on the truncated Cavan run having been cancelled and the land opened up to a swathe of new selectors. Clearly, competition for land in the area had become intense and there was enormous pressure on the Roches to consolidate the Cavan estate by acquiring as many blocks under the conditional purchase scheme as possible to prevent new settlers coming into the area. Between 1891 and 1893, there are innumerable references in the Cavan diaries to Eliza and Fred making court appearances in Yass to protect their landholding and keep out new selectors. One diary entry dated 31 July 1890 describes how Fred and Eliza went into town to try and secure four lots on the Cavan run and that there were no fewer than forty other selectors present on what was the opening day of the remaining leasehold area of the Cavan run being thrown open for selection.

Two valuable blocks that the Roches failed to secure were on the river flats originally purchased by Thatcher as a dummy on behalf of Castle and which were bought by John Styles in the early 1900s. Styles was an obvious buyer as he already owned an adjoining block on the river to the south and a number of other blocks that ran along the banks of the Murrumbidgee on Cavan's eastern boundary. Since these river flats were extremely covetable, the most plausible explanation for their 'loss' to Styles is that the Roches could simply not afford the final balance due on the conditional purchase, given all the other calls on their capital.

Whilst the Roches may initially have been upset or disappointed over failing to secure these blocks, relations at Cavan between the Roches and Styles are not known to have been strained. On the contrary, Styles was subsequently employed at Cavan as a building contractor and was responsible for the woolshed at Three Oaks, completed in 1899.[15] Known for his fiery temper that earned him the nickname of 'Hellfire Jack', he was 'universally esteemed', settling on his property by the river and calling it Wynburn.[16] Born into a staunch family of Protestants, he had married, after much opposition from his parents, Margaret Booth, a Roman Catholic. According to family tradition, the couple remained sweethearts until he died in 1924. Much later in the history of Cavan, Styles's grandson, Wilfrid Wade, was to marry into the Castle-Roche family, an indication perhaps that whatever disappointment had occurred in the past as a result of Styles 'poaching' some valuable river flats on the Cavan run, it was now forgotten.

PROGRESS AT CAVAN

By 1897, the Roches had managed to acquire thirty-three conditional purchases of land covering 7,000 acres, but this required funds not only to purchase the land itself, but also to fulfil the government requirements to improve it with, amongst other things, fencing and water facilities for stock. No doubt in an effort to generate more income, Fred increased his stock numbers again to just over 12,000 sheep. Even though Cavan was by now only 12,603 acres, it was a gamble that seems to have paid off and saw the property safely through the turn of the century. In the first decade of the 1900s, sheep numbers were reduced again to between 8,000 and 9,000 on a steady landholding of just above 12,000 acres in total.

Throughout the stewardship of the Roches, the main focus on

By 1897, the Roches had managed to acquire thirty-three conditional purchases of land covering 7,000 acres, but this required funds not only to purchase the land itself, but also to fulfil the government requirements to improve it with, amongst other things, fencing and water facilities for stock.

—

Cavan was very much wool production, and there is no longer any mention in the diaries of any horticultural activities nor of Castle's vineyard. Unfortunately, the diaries do not give the type of sheep that were being grazed on Cavan, only details of the wool clip, which increased steadily each year. Like Castle before him, Fred had a strong workforce of six or seven men working on the place, including his brother, Andrew, from time to time. There was also Laddie when he was home from school for the holidays, and there are several diary entries indicating that father and son rode out together regularly. Naturally, the young boy was being groomed to take over from his father one day and seemed to be happy in the process.

Luckily, from a farming perspective, in the late 1800s and early 1900s there appear to have been no major calamities at Cavan, other than an extensive fire that raged for several days all the way up the Murrumbidgee from Gundagai towards the south, but which only came as far as the Cavan gate, burning out a single paddock called Wade. In fact, the diaries are remarkable for recording only one significant fire in the district, in spite of some severe periods of dry, such as in mid-winter 1907. On

There is scant evidence that Fred and Eliza had much of a social life, possibly due in part to Eliza's elderly mother who was still living at Cavan and to Fred's workaholic tendencies.

—

30 July of that year Fred wrote, 'This section of the river valley has never within our recollections looked so wretched as during this winter; there has been no growth and even within our wire netted enclosure the flowers and vegetables have failed to respond when watered.'

That 'wretched' winter was in stark contrast to the one at the end of 1890, when Fred recorded, 'This year has been a glorious one for grass and water. Two remarkable features during this period were that on the 9 and 10 August, the heaviest fall of snow experienced for many years fell on the river and the almost total absence of frost all the winter. The spring was very late in comparison to previous years.' It seems that thirty inches of rain fell that year, two or three inches above the average in the district at the time. Good rains could be a mixed blessing, however; in June 1891, there was a big flood at Cavan that reached a height just below the great Gundagai flood of 1852. Fred wrote in his diaries that the water measured two feet up the walls of the homestead which, at that time, still stood close to the river on a sweeping bend, and came within three feet of the top of the old Taemas Bridge, causing traffic to be suspended for twelve hours.

Other than the typical amount of hard, physical work required on such a property, life at Cavan seems to have been very quiet in this era, especially after Laddie went away to school. There is scant evidence that Fred and Eliza had much of a social life, possibly due in part to Eliza's elderly mother who was still living at Cavan and to Fred's workaholic tendencies. The dark clouds of Eliza's mystery illness that began in earnest in the early 1900s might also have been looming. However, the Cavan diaries do mention regular visits by Andrew Roche and his family and by Eliza and Fred to Ravenswood. In particular, the two daughters of Andrew and Harriet Roche were frequent guests at Cavan in the summer holidays after Christmas. Nevertheless, things began to change soon after Wilhelmina Castle died on 6 August 1899 at the age of ninety. Within six months of her death, plans for a new and very elaborate homestead were drawn up, perhaps in anticipation of a different kind of life for its inhabitants from the one led there before.

THE NEW HOMESTEAD

On her death,[17] Wilhelmina Castle left her half share of Cavan, inherited from her husband, to her nephew, Fred, for his lifetime. In a curious twist to the Cavan story, this bequest was made in a will dated 11 January 1884,[18] three years before Fred married Eliza and became Wilhelmina's son-in-law. It seems unlikely that Wilhelmina meant to undermine her daughter's position at Cavan by this manoeuvre and the most plausible explanation is either that Eliza was already engaged to Fred at the time the will was drawn up, or that Wilhelmina was anxious to facilitate a union between the two cousins by making Fred a man of property.

The estate, which also included two small parcels of land at Wollongong, was valued at 'under £5,920' and said to be subject to various mortgages, half of which Fred was obliged to

Top — *Unknown figures beside Grecian portico, Cavan homestead, 1906*
Bottom — *Cavan homestead veranda; Laddie second from left, his parents Fred standing on his left and Eliza seated front right, c. 1908*

Top — *Cavan homestead on completion, c. 1906*
Bottom — *Cavan shearing shed, 1908*

pay off whilst Eliza was said to be responsible for the other half. Undoubtedly, Cavan was worth less on Wilhelmina's death than it had been on the death of her husband, as it was now a fraction of the size and borrowings may also have increased to finance conditional purchases. Land values had also fallen during the depression of the 1890s. Presumably, in anticipation of Eliza and Fred becoming tenants in common of Cavan on Wilhelmina's death, many of the conditional purchases of land made to consolidate the property had been taken up in both their names.

Whilst the worth of Wilhelmina's estate was modest compared to that of her husband, it still represented a considerable sum at the time. Part of it may have been the means by which Eliza and Fred were able to afford a new house on a grand scale and with some unusual elements. Joseph Alexander Kethel, a Scotsman from Perth and a prominent, award-winning architect of the day, was commissioned to draw up plans for a homestead.[19] According to the Cavan diaries, it took Eliza and Fred just five days to agree to most of the design, and by 26 January trees were already being cleared on the new site, a hill about half a mile further away from the Murrumbidgee than the existing limestone homestead. After the flood of 1891, the couple no doubt took the precaution of moving onto higher ground further from the river.

From this moment, Fred's attention seems very much to have swung from his usual farming activities to supervising progress on the new building. The Cavan diaries are full of entries alluding to the building works. Touchingly, Andrew Roche and his wife, Harriet, were asked on 14 February 1900 to lay the foundation stone of the new homestead, 'placing a bottle with the usual items under it'. Kethel, who had designed several public and commercial buildings in Sydney, including the Sun building in Castlereagh Street and the Alliance Assurance Company's

building in Pitt Street, made regular visits to Cavan throughout 1900 and 1901 to inspect the works.

The house was constructed of limestone bricks with lavish interiors of pressed metal Wunderlich ceilings and hand-painted glass panels set into wooden doors. Most noteworthy of all was an extraordinary portico in the style of a Greek temple, supported by white painted timber columns, that was built alongside the house at the front entrance. Whether this was an idea of Kethel's or an affectation on Eliza's part – a nod perhaps to her late father with his interest in classical civilisation – is not known. Certainly, its construction and completion caused more problems than any other part of the project, and the painting and repainting of its columns and pedestals is frequently mentioned in the Cavan diaries.

Curiously, work on the new homestead came to an abrupt halt in 1902, a year of very dry weather. Sheep numbers on Cavan had risen in the previous couple of years to over 14,000, although the acreage was still steady at about 12,500. It is tempting to deduce that Fred had increased his flock to help pay for the new homestead but then came under pressure during a low-rainfall period because of this overstocking and a shortage of feed. His halving the flock the following year to just 7,546 supports this analysis. A diary entry of 12 January 1903 reads: 'The river is so dry it can be crossed almost anywhere dry shod.' Fortunately, the cash flow crisis was over by 1904 and work recommenced on the homestead, continuing until about the middle of 1906. Once it was finished, the homestead was a handsome but somewhat severe two-storey building with polished timber floors, a long veranda and a glass conservatory. In keeping with the new fashion, the kitchen was incorporated into the house rather than situated in a separate building with the servants' quarters as had been the traditional style previously.

— Eliza Castle-Roche, John Vickers, Jessie Hill, Wilhelmina Castle and Laddie Castle-Roche at Cavan, c. 1895

It is in this last stage of completion that references to Eliza's ill health first surface. It seems that she went on a visit to England at some time in the first half of 1905, partly or wholly as a remedy for debilitating headaches that she was suffering regularly. In a letter of 2 September 1905, Fred says that he 'was pleased to hear you seem better in the upper mechanism – it would be a great boon if the headaches forgot to return'.[20] Surely Eliza would not have gone abroad during the final phase of building the homestead unless she had been unwell for some time and was perhaps also seeking medical attention beyond her usual doctors at home? The cause of the headaches does not seem to have ever been ascertained until her death; but there is also reference in Fred's letters to Eliza's low spirits, which may suggest that at this stage she was suffering from depression, either by itself or in conjunction with some other sickness. However, shortly after her return from England with a close friend, Caroline Crosby, Eliza was well enough to attend Laddie's entry into Sydney University in March 1906 and to shop for vases for the Grecian porch, which she had already adorned with statues purchased whilst in London. It was also at this time that the family graves at Cavan containing Eliza's parents were opened, their remains placed in the same coffin and moved to a new vault behind the modern homestead.[21]

ELIZA'S UNTIMELY DEATH

But Eliza was still unwell and her eyes especially were giving her trouble, so she returned to Sydney in March 1909 to see her doctor. Several loving letters from Fred in this period survive. He was clearly worried and told Eliza to instruct Laddie to help her with her correspondence. Fred could not get away from Cavan to visit Eliza due to a shortage of staff around the homestead. He was, nonetheless, constantly reassuring that all was well at home – 'we are jogging along in the same old way' – and full of news of day-to-day life on the property. In relation to Eliza's sickness, Fred seemed at a loss as to what to say after what appears to have been at least six years of suffering and no proper diagnosis. He suggested a spell on the coast either with Laddie or a friend, saying in a letter of 27 March 1909, 'it would set you up again and be a rest and perhaps tone the system, you might ask the Dr what he thinks of it as regards the effect it would have on the eyes and head'. The doctor's advice is not known, but Eliza went to Brisbane in early May to stay with the Crosbys. Two months later, she was dead. The chief cause stated on her death certificate is a cerebral tumour, the first mention of this in any surviving records.

The last letter from Fred to his wife (undated but about the end of May) makes heartfelt reading. Addressing Eliza as usual as 'Best Beloved', he says, 'Our one wish is to see you return better and eventually become strong again.' With that, Fred enclosed news of the engagement of their niece, Marguerite Daisy Roche, Andrew and Harriet's daughter, and news of Laddie's great success in fixing and driving a new engine at Cavan. Never does there seem to have been any concern amongst the family that Eliza was in real danger until she arrived in Queensland. There, the doctor diagnosed 'congestion of the nerves of the brain', aggravated by her suffering from dengue fever, apparently contracted on her voyage from Sydney. She was moved to a private hospital on 8 June 1909 and Fred rushed away from Cavan the same day.

Arriving in Brisbane two days later, Fred had ten days to be with Eliza before she died at 2.15am on 21 June. She was only fifty-nine years old. According to Fred's entry in the Cavan

The last letter from Fred to his wife makes heartfelt reading. Addressing Eliza as usual as 'Best Beloved', he says, 'Our one wish is to see you return better and eventually become strong again.'

—

— *Hunting party with Laddie (centre) and his father (right), 1908*

— *Cavan cottage with unknown figures, c. 1908*

On Eliza's death, her half share of Cavan was valued at three times the amount her father had left her. Eliza and Fred had enhanced the value of Cavan dramatically with astute land purchases and improvements combined with skilled management.

—

diaries, 'her end was peaceful and practically painless'. Fred travelled home with the coffin, reaching Cavan on 24 June and burying Eliza in the family vault that afternoon. As Caroline Crosby wrote movingly to Fred on 9 July, 'We thought of you and your son at the vault. The rain must have made it a desolate scene. I hope you had someone to comfort you after you must have sorely needed it after all your trying experience here on the way and the final laying to rest. Then the empty place in the home ...'[22]

The only obituary for Eliza, which appeared in the *Yass Courier* the day after her burial, is short and to the point. No mention is made of her personality or any contribution she may have made to the local or wider community. Perhaps she had always been a very private, retiring sort of woman not given to attracting any attention. Nonetheless, as Fred's companion for more than twenty years, he must have been desperately lonely after her sudden death, even if he was a man of the land, used to working long and sometimes solitary hours amongst his sheep. A one-line entry in the Cavan diaries a year later, on 21 June 1910, says it all: 'Anniversary of Sweetheart's passing.'

It seems that Laddie at least came home to stay permanently at the end of 1909, which would have been some consolation for Fred. The sole beneficiary under his mother's will, Laddie had now become co-owner of Cavan with his father.[23] On Eliza's death, her half share of Cavan, together with livestock and bank deposits of £2,500, was valued at the significant amount of £20,683, three times the amount her father had left her. Without question, Eliza and Fred had enhanced the value of Cavan dramatically with astute land purchases and improvements combined with skilled management.

THE NEXT GENERATION

According to the Cavan diaries, there was little interruption in the usual round of activities on the place in spite of the tragedy of Eliza's untimely death. A few more uneventful years passed and the new homestead must have been a cruel reminder of plans for the future snatched away. Almost as suddenly as his late wife, Fred was taken dangerously ill during shearing in October 1915. He was operated on in Yass Hospital for some obstruction in his intestines. The operation was unsuccessful and he was re-operated on two months later. Laddie was summoned to his father's bedside when a heart complication arose, but nothing could save him. Fred died on Christmas Eve 1915 at the age of sixty-one. Laddie wrote in the Cavan diaries, 'Father passed peacefully away at 11.05am being conscious to the last cause being shock and heart failure after anaesthetic on Monday last.' Laddie, at only twenty-seven, had now inherited the whole of Cavan.

When Fred died, the obituary in the *Yass Courier* on 30 December 1915 spoke of his integrity, calling him 'a man whose word was, if possible, more than his bond' and referred to his 'exceptionally liberal' donations to deserving causes (though these are not enumerated). Fred left an estate valued at £14,541 and had the satisfaction of knowing that the property – whilst now only about 12,000 acres and under half the size it had been when he married Eliza – was nonetheless, through his own hard work and dedication, far more valuable, compact and reliable than before. In his time at Cavan, the less productive country previously held under lease had been relinquished and the choicest land taken up by himself and Eliza under the conditional purchase scheme. Fred would also have known when he died that his son was cast from the same mould as he with a similar enthusiasm for hard work and that he would continue to build upon his parents' achievements, which is exactly what he did.

— *Men repairing machinery at Cavan, c. 1908*

CHAPTER TEN

THE CASTLE-ROCHE YEARS

LADDIE HAD PRIMARILY BEEN EDUCATED at King's College, Goulburn, not far from Cavan. The school was founded in 1888 by a former master of Scotch College in Melbourne and attained a high degree of prominence for the excellence of the education offered. Nothing is known of Laddie's school career other than that he passed his matriculation in March 1906,[1] whereupon he went to Sydney University to study geology, registering simply under the surname of 'Roche'.[2] Having grown up amongst the Cavan limestones, it is not surprising that Laddie was interested in rocks, but it seems he never graduated, although he passed his examinations in the first two years. Instead, with the sudden death of his mother, Laddie chose to return to Cavan to manage the property with his father.[3] About this time, according to the Cavan diaries, he also attended a technical college where he followed a course in wool classing and acquired a skill in mechanics.

On his father's death in 1915, Laddie inherited one of the oldest, most prestigious properties in New South Wales, one that had been in his family for two generations. He must have been at once excited and a little daunted at the task ahead. Cavan had a history that demanded respect and careful preservation, but fortunately Laddie had been well prepared, having spent his entire life in Cavan's awesome and sometimes hostile landscape. By the time he succeeded his father, he would have known every square inch of the property, every tree and every rock, and been grateful to have learnt so many valuable farming lessons at Fred's knee. Since he was also an indefatigable worker, Laddie was as suited to the management of Cavan as any young man could be and exuded a quiet confidence that made him a leader amongst his workforce as well as popular with his many friends.

Above — *William Frederic Fletcher Castle-Roche (Laddie), c. 1910*
Previous pages — *Cavan wool going to Sydney, December 1922*

By now, Cavan had been steadily improved with the acquisition of the better grazing lands in the area and the disposal of the more inferior ones. Good management had ensured that a rich variety of grasses and herbage flourished on the hills, including trefoil, crowsfoot, barley grass, rye grass and subterranean clover; and on the river flats, there was an abundance of lucerne, cut and laid by for use in the drier years. There was also plenty of permanent water for stock, not only because of the usually reliable Murrumbidgee River, but also due to the numerous mountain streams and creeks and the fairly generous average rainfall of about 28 inches a year. Without question, all of these features boded well for a successful transition to the third generation and a bright future for Cavan.

IMPROVEMENTS

Although Cavan was much developed by this time, it was also overrun with rabbits, a common problem in this era. Laddie was determined to rid Cavan of this problem and every effort was made over several years to rabbit proof the property, paddock by paddock.[4] Fences were rigorously netted and gates rebuilt, then under-laid with netting and logs so that rabbits could not creep beneath. Rabbit burrows were then dug out in a strenuous effort to eliminate them entirely and reduce the chance of re-infestation. Stock theft was also not uncommon at this time and the new padlocked gates helped to keep this risk to a minimum. Two men were constantly employed to patrol the paddocks just to keep the rabbits at bay and to eliminate weeds. Whilst tedious and costly, these measures greatly increased the carrying capacity of the land, possibly by as much as one hundred percent.

As a major capital improvement, Laddie designed and built

— *Machinery shed, c. 1930*

Cavan had been improved with the acquisition of the better grazing lands in the area and the disposal of the more inferior ones. Good management had ensured a rich variety of grasses flourished on the hills, and there was plenty of permanent water for stock. All boded well for a bright future for Cavan.

—

179

— Vehicles parked outside original machinery shed today

in the 1920s a massive complex of stables and a workshop that still stands today. This building – which, like the homestead, was constructed of local limestone – had walls eighteen inches thick and contained seven stalls, a harness room, feed room, meat chamber and machinery store. From 1937, it was lit throughout by electricity. Inside, the workshop, which had a large carpenter's bench, was equipped with a circular saw, lathe and drill press, all connected by belts to a common overhead driveshaft powered by a Lister engine. The saw was used for cutting timber for gates, fencing and framing timbers necessary in any building works on cottages and sheds. This was an important advance in terms of efficiency on the property as, formerly, all timber had been sawn by hand. In a separate annexe of the building, there was a paint shop and a blacksmith's shop complete with the usual forge and anvil. At this time, there were about twenty-seven working horses on Cavan, including a stallion and three draft horses, plus sundry mares and foals.

Laddie was very conscious of soil erosion through over-stocking and also strongly believed in pasture improvement, under-sowing his crops of wheat, oats, corn and lucerne with subterranean clover and phalaris. Although crops were grown in this period, they were mainly for station use; Cavan continued to be essentially a wool-growing property, carrying about 13,500 merinos of a similar fine wool variety that had made the property famous in the past. However, the wool was superior to that of the original Riley Saxons, as breeding techniques had evolved and refined the fleece. Rams were purchased that had Merryville bloodlines, a gene pool which owed its origins in part to the Rawdon stud founded on Riley Saxon sheep purchased by Edward Cox after the death of William Riley.[5] The Merryville sheep were larger and more plain-bodied (less wrinkled on the neck and shoulder) than the original Riley Saxons, having been constantly 'improved' through breeding to produce a longer, stronger staple and finer, heavier fleece. At the time, these sheep were the aristocrats of the fine wool type, and even today the Merryville stud occupies a pre-eminent position in the industry. In later years, some of the merinos at Cavan were crossed with Romney Marsh and Southdown sheep for fat lamb production, but wool growing remained the priority. It appears that in

Laddie was a well-respected employer, not only because he himself worked very hard, but also because he was kind, sincere, loyal, generous and extremely modest. He also had a keen sense of humour and a hearty laugh.

—

contrast to the Riley era, Cavan wool was largely sold into the domestic market and during the First World War, much of the clip was sold to Vicars Woollen Mills in Sydney to make army uniforms.

THE WORKFORCE

Laddie was a well-respected employer, not only because he himself worked very hard, but also because he was kind, sincere, loyal, generous and extremely modest. He also had a keen sense of humour and a hearty laugh. These qualities naturally endeared him to his workforce, comprising at least five men at any one time, most of whom stayed for many years. According to his daughter, Lorna, a man called Bill Noyes worked for three generations of the family, so he must have arrived at Cavan in the Joseph Castle years. He lived in a cottage in the area now known as Cavan West, which he managed and maintained almost by himself. Every evening, he would report by telephone to Laddie and plan the work for the next day. Henry Harmer was another long-term employee who arrived at Cavan in the late 1930s. He was a single man with very good mechanical and building skills and lived in a tent, which he pitched wherever he was working at the time. According to Lorna, he seemed to prefer it that way!

Above all, however, Laddie valued James ('Jim') Northey, who arrived at Cavan as a station hand with his wife and three daughters on 2 June 1939.[6] Laddie had met Jim at Eumbra, another smaller property at Barraba in northern New South Wales, which Laddie had purchased in 1925[7] and where Jim's brother was overseer. Jim was an exceptional man, the epitome of the conscientious and dutiful farmhand, working at least ten hours per day for six days one week and five and a half days the

Top — *Jill Northey with Flossie at the Yass Show, c. 1947*
Above — *Jim Northey chaff cutting at Cavan, c. 1950*

next. He was one of seventeen children apparently born in a tent on the banks of a river somewhere in the vast expanse of the Australian bush.

At Cavan, Jim started off by looking after the cropping, supervising the wheat harvest and making chaff out of the stalks. He was also the blacksmith and farrier as well as being responsible for water on the property, maintaining the big tank on the top of the hill next to the family vault and keeping the windmills in good order so that there was always water for stock in the troughs. Jim and his wife, Ivy, and their daughters, Lorraine, Jill and Roz, lived in a simple cottage on the place, also built of limestone. The house was dark and cold in winter with only one fuel stove and an open fire place in the kitchen and sitting room. The kitchen was the only place with electricity and all laundry was ironed with flat irons heated on the stove. The lavatory was a 'long drop' and some distance from the house, up past the wood pile. The family was very poor, with Jim earning only £3 per week, aside from his meat and milk rations and the provision of the cottage. His daughter, Roz, recalls ruefully that she was seventeen years old before she was given a new dress.

THE ABERDEEN ANGUS CATTLE STUD

A further valuable addition to the workforce was Wilfrid Wade who joined the team at Cavan about 1928.[8] He had grown up on the next-door property, Wynburn, owned by his maternal grandfather, John Styles, the firebrand who had acquired the best river flats to the east of Cavan during the early 1900s and become a successful, widely respected grazier in the Yass district. His paternal grandfather, Abraham Wade, had also been a prominent pastoralist in the area, having owned Taemas and Salt

— *Cattle drifting down track to original Cavan homestead site*

Box stations,[9] the latter property now part of modern Cavan. According to Castle-Roche family history, Wilf ('Winkie' or 'Wink' as he was also known) had always had a close, almost fraternal relationship with the older Laddie, in spite of the fifteen-year age difference between them.

Wilf was predominantly interested in cattle and persuaded Laddie to go into partnership with him to start an Aberdeen Angus cattle stud which was run parallel to the sheep operation. Wilf was barely thirty years old at the time, and although his family was well off, there is no record of whether or not he contributed to the capital of this venture. In any event, the few shorthorn cattle that were still at Cavan were sold and Wilf went to New Zealand in 1934 to buy five females and one bull from the famous Gwavas stud.[10] These animals played a seminal role in the development of what became a first-class Angus stud at Cavan and which was further enhanced by the purchase one year later of two more females from the Waiterenui stud, also in New Zealand. The aim was to breed an early-maturing type of animal, low set, with deep barrel, which would produce beef more efficiently and bulls suitable for export. The herd, apparently not affected by the long cold spells in winter, thrived at Cavan, and when Wilf went to war Laddie's wife, Enid, took over the running of the stud.

FAMILY

On 6 April 1921 at St Philip's Church in Sydney, Laddie had married Enid Amy Fox, the twenty-year-old daughter of the late George Fox, a well-known Queensland pastoralist who, prior to his decease, owned properties near Rockhampton where Enid was born. Her mother, Amy, was the daughter of the late John Badgery, a pioneer of the pastoral industry of New South Wales and famous for making several epic overlanding trips, droving cattle from the slopes of Mount Kosciuszko right out to western Queensland.[11]

Laddie met Enid at Bloomfield, the charming property just north of Cavan over the river that was once owned by the ex-convict George Davis, and which is now part of the Cavan estate. At the time the couple met, it was a soldier's settlement block owned by Malcolm Badgery, whose half-sister, Amy Fox, was Enid's mother. Amy Fox was at that time widowed and, together with her three daughters, lived with her half-brother who was in poor health after the First World War. When the legendary John Badgery died in his eightieth year in 1916, his estate was not free from debt; his property, Burra, south of Queanbeyan, that had been left to Malcolm and his younger brother, Ronald, had been sold by the executors, which compelled both brothers to make a life elsewhere. Amy too had an inheritance from her late husband and had purchased Warroo, the property adjoining Bloomfield that had previously belonged to William Riley. Warroo did not have a house on it, but as Amy lived with her brother, she had a home and he had a housekeeper.

Laddie and Enid's wedding appears to have been a modest, largely family affair as there are no wedding photos that have survived and no reports have been found in the local or Sydney press. The couple had two children, William Frederic Malcolm (known as 'Bill') born on 2 April 1922, and Lorna Brenda born on 11 July 1924. Bill was sent away locally to school at the age of only five and then later to Barker College in Sydney where, amongst other things, he received instruction in wool classing. Lorna was educated by her mother at home until she was ten years old whereupon she was sent to Wenona School in Sydney.

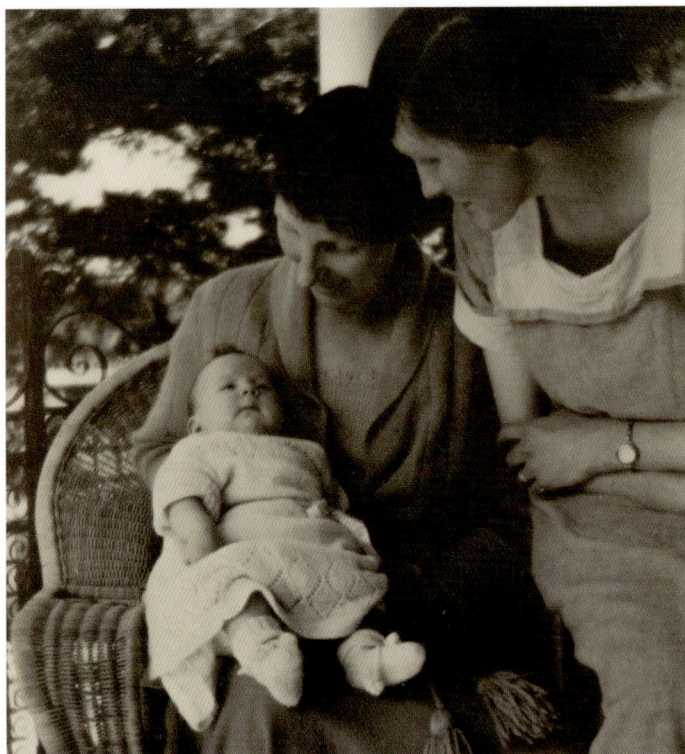

Above top — *Amy Fox, Enid Castle-Roche and her son Bill, c. 1922*
Above — *Bill Castle-Roche, c. 1928*

Enid was a good, kind mother, devoted to homemaking and family life; and she was much respected by her husband to whom she was happily married until he died.

Lorna has wonderful memories of her childhood at Cavan and talks about the magical garden – a wide lawn surrounded by a profusion of flowers, including a mass of roses in vibrant colours. Many of these specimens came from cuttings taken from the old homestead garden that still flourished down by the river below a steep hillside of kurrajong trees. Here was also a source of walnuts, apricots, plums, figs, mulberries and nectarines, a Garden of Eden that produced so much fruit that the surplus was bottled or made into jam to avoid wastage. Enid spent a great deal of time in this garden and had even found asparagus growing there, possibly from the plants first established by James Calvert and Joseph Castle. The vegetable garden of the new homestead became a beneficiary of these old asparagus beds and was a very large area, tended to diligently by Enid and a full-time gardener. Both her parents were devoted to Cavan, Laddie to the farming and Enid to her gardens.

Lorna remembers that life at Cavan was very quiet. In those days, there was little social life beyond tennis parties and family picnics, a contrast to the more flamboyant years that came later. Lorna amused herself with sewing, gardening and a little riding, usually only if there was a muster in progress and her help was needed. Swimming in the river was out of bounds as the cold, inky black water was treacherous with a strong current. However, there were pets and an area below the garden known as 'the Park', with a high fence that became home to generations of pet lambs and kangaroos. Apparently, there were few kangaroos on Cavan in this period, although there was a mob that could generally be seen grazing in the wooded areas of Narrangullen next door.

— Feeding silage in 1940 drought. Lorna Castle-Roche is standing against the truck to the left

THE FLOOD OF 1925

On 25 May 1925, the greatest rainfall in the living memory of white settlement in the Yass district began and continued for several days.[12] The Yass River rose thirty-five feet above its summer level and the Murrumbidgee swelled to such an extent that it was two miles wide in some places. People lost their homes and their livelihoods and the raging water washed away the Taemas Bridge.[13] A line of trees, logs, haystacks and general debris floated down the river during the flood, lodging against the bridge and causing water to build up behind it. The extreme force pushed the steel spans off the pylons and caused the bridge to collapse in its entirety, only twelve months after it had been raised as part of the Burrinjuck Dam storage scheme.

The original Cavan homestead down by the river was also engulfed, leaving only the outer walls of the house and some veranda posts, clad in a heavy and winding wisteria. A pergola covered in a sprawling banksia rose also survived, and in later years these and other remnants of the garden crept over the ground to climb nearby elms and fruit trees, making a glorious scented canopy. As for the new homestead, the flood water came right up to the garden fence, submerging the drive beneath, but otherwise, there was no damage within the curtilage of the house and grounds. The Bloomfield homestead, which was built perilously close to the Murrumbidgee on the flat by the Taemas Bridge, was inundated and damaged, but miraculously survived.

With the loss of the Taemas Bridge, a punt was established,[14] harking back to the 1850s when this was the only method of crossing the river, other than swimming or riding a horse further upstream over the Cavan or Bloomfield fords. In those early

Top left — *Former blacksmiths shop destroyed by 1925 flood*
Top right — *Original Cavan homestead flooded in 1925*
Centre — *Murrumbidgee in flood, 1925, looking east from Taemas*
Bottom left — *Stables destroyed by 1925 flood*
Bottom right — *Former men's quarters after 1925 flood*

days, an enterprising settler at Taemas, Richard Gorman, had seen the need for some form of permanent crossing with the increased traffic of settlers and later gold miners, heading to the mountain areas around Kiandra.[15] He therefore stretched a rope across the river, built a raft that could carry one ton in weight and went into business transporting people, animals and a wide variety of goods and chattels. In similar fashion after the flood in 1925, another punt was introduced that ran on a cable propelled manually by a punt man, this 'modern-day' Charon being summoned peremptorily from his nearby cottage by the blowing of a car horn. For six years, this hopelessly old-fashioned arrangement prevailed until the new Taemas Bridge was finally erected.

— Punt in place of old Taemas Bridge, c. 1930

THE NEW TAEMAS BRIDGE

The first Taemas Bridge had been built in 1887, after local settlers had lobbied for more than twenty years for a safe and reliable crossing over the Murrumbidgee in order to access the road to Yass. The opening of the railway at Yass Junction in 1876 meant that farmers were assured of direct transport to Sydney for their produce, if only they could get it there securely. It was Cavan's far-sighted occupant, James Calvert, who put forward the first proposal for a bridge at Taemas, but it was not until 1882 that the government agreed to finance the scheme.[16] Tenders were advertised in 1884 and work began the following year, with the bridge being completed on 30 November 1887. The bridge had cost £10,384 and consisted of three wrought iron lattice girders, the central span being 182 feet and the two side spans 140 feet each, all supported on two abutments of local limestone and two cylindrical piers filled with concrete. A

formal opening ceremony was held on 14 April 1888; the event was considered so important that the day was declared a public holiday in Yass.

With the construction of the Burrinjuck Dam some years later, the water level of the river rose and the bridge was too low to cope. It was therefore raised fifteen feet and also extended on the sides, but this proved insufficient during the flood of 1925. After several attempts to find a solution, the new Taemas Bridge that still stands today was erected two miles upstream from the old bridge, occasioning a longer road route to Yass, but where a shorter bridge could be constructed and flood levels were expected to be lower. This bridge was completed in 1931 at a cost of £60,895 and in spite of several major floods since, stands proud, high above the river between two hills, linking the original Cavan grant to Wyelba, another smaller property now included in the Cavan estate.

Opposite
Top left —*Battery room, c. 1930*
Top right — *Back of Cavan homestead with old kitchen and cool room, c. 1930*
Bottom left — *Cavan homestead, c. 1930*
Bottom right — *Cavan homestead hall, 1908*

This page
Top left — *Hallway at Cavan with hand-painted panelled doors, c. 1930*
Top right — *Dining room at Cavan, c. 1930*
Left — *Cavan homestead drawing room, c. 1930*

THE FIRE OF 1939

The flood of 1925 was not the only calamity of the Castle-Roche era. On the evening of 12 January 1939, the night before what became known as Black Friday, a small dinner party was in progress at Cavan. About 8pm, a phone call came through seeking help to fight a bushfire out towards Tumut behind Wee Jasper. The male dinner guests were fitted out with work clothes and joined hastily gathered station hands before setting off for the fire. Cavan and Narrangullen were the only properties in the area with mobile water tanks. Cavan had a three-ton truck with a ship's water tank on board. Each man also wore a water container on his back and carried beaters, rakes or axes. Apparently, Enid did not go to bed that night and did the washing to take her mind off the drama. She was found by Lorna to still be ironing the next morning at 5am, the washing having dried in the high temperatures of the evening.

Most of the men stayed throughout the night and the next day fighting the fire. The heat and wind made it impossible to control and all they could do was to keep people and buildings as safe as possible. According to the diary of the young Bill Castle-Roche,[17] temperatures had hovered around 41 degrees Celsius a few days before the fire began and continued at this level during the course of that weekend. Bill Noyes telephoned Enid from Cavan West on the Friday morning to warn her that he thought the Cavan homestead and outbuildings would go up in flames. The house was surrounded by very flammable pine trees and limestone would burn. 'Be prepared but do not try to save it', was Bill's advice. Luckily, the wind changed direction, driving the fire further south and taking the danger with it.

Nevertheless, the fires were the worst in the history of the Yass district, destroying over 115,000 acres of grass, 11,350 sheep and at least 270 miles of fencing.[18] Narrangullen, next door to Cavan, was burnt out completely, the homestead being narrowly saved by a patch of lucerne and a wall of bricks ready for the construction of a new building. Goodradigbee was also devastated; 35,000 acres were lost along with 5,000 sheep, fifty miles of fencing, sixty tons of hay, a woolshed, stables and yards. At Cavan, although no buildings were lost, only 1,500 acres of grass were saved out of about 12,000. Stock losses were huge and at least ten miles of fencing was destroyed, with few fences remaining to contain sheep and cattle that survived. For weeks afterwards, men out riding carried rifles strapped to their saddles in order to destroy animals with burnt feet or smouldering fleeces, saving them from a painful and lingering death.[19] All healthy sheep were gathered up and sent for twelve months to a property leased at Umbango Creek in Tarcutta about one hundred miles south. According to Bill's diary, his father had wanted to send the sheep to his property, Eumbra, in northern New South Wales, but there was not the pasture there for this to be feasible.

> The fires were the worst in the history of the Yass district. At Cavan, although no buildings were lost, only 1,500 acres of grass were saved out of about 12,000. Stock losses were huge and at least ten miles of fencing was destroyed, with few fences remaining to contain sheep and cattle that survived.
>
> ——

In addition, Bill noted that his father was very worried about all the fencing that was lost, about £8,000 worth, and which took two years to replace. This was done as quickly as possible, but the fire had destroyed many trees on Cavan so that reinforced concrete and steel posts had to be used to supplement the wooden ones that were usually made from timber cut on the property. The costs associated with the fire – agistment, fencing, restocking and loss of income – caused Laddie immense anxiety and created an enormous debt burden. As if this was not trial enough, the Second World War had broken out and manpower was becoming extremely short.

> Bill was posted variously to Palestine, Egypt, Lebanon and Syria, though it is unclear whether he saw action in any of these places. Certainly, he was at the second battle of El Alamein and was so shaken by his war experiences, he was scarred ever after. During his absence, his father wrote frequently to Bill with news from home.

—

THE WAR YEARS

Wilfrid Wade was one of the first to join up. He had a great sense of duty to his country, matched only by his loyalty to and affection for Laddie and Enid, as is evident from the many letters that he wrote to them whilst he was fighting the Japanese. In one letter written from Malaya on 20 December 1941,[20] he said, 'The Jap is having his run of luck just now but it won't last. It would do you good to see our boys burnt black and as hard as nails, just waiting to go straight into him. You can't beat these Aussies for settling down to all sorts of conditions and making the best of them ... don't worry we will do our job.' Wilf was subsequently taken prisoner and languished for two years in Changi prison camp, Singapore.

When he left for Malaya, Enid, who had always been interested in cattle, took on Wilf's role as manager of the Aberdeen Angus stud, assisted by Ivy Northey.[21] The stud cattle were mostly kept in two paddocks, one close to the homestead and the other about two miles away where the cows were grazed.

When not in use, the stud sire, called 'Grantham', was kept close to the house where he apparently enjoyed the quieter life away from his harem!

Whilst Wilf was away, Laddie, who was too old to sign up, remained at Cavan and concentrated on sheep work. His son, Bill, joined the army and sailed from Fremantle on 7 September 1942 for the Middle East. According to his leather-bound notebook,[22] Bill was posted variously to Palestine, Egypt, Lebanon and Syria, though it is unclear whether he saw action in any of these places. Certainly, he was at the second battle of El Alamein and was so shaken by his war experiences, he was scarred ever after. During his absence, his father wrote frequently to Bill with news from home.[23] Describing the particularly good spring they were having with grass that was almost too much for the stock to cope with, he lamented the shortage of labour, which meant that Enid had become 'the spare part that fits everywhere', dipping sheep and driving machinery.

Above top — (Back, left to right) Enid Castle-Roche, Lorna Castle-Roche, Amy Fox, (front, left to right) Bill Castle-Roche, Wilfred Wade and Laddie Castle-Roche; 1940
Above — Bill Castle-Roche and his father, 1940

The war went on, far away from the mostly female inhabitants of Cavan, who were nevertheless issued with guns, hand grenades and gas masks and given coupons for tea, sugar, clothes and petrol. The reliable Jim constructed a dugout just below the tennis court where the occupants of Cavan could hide if an invasion occurred. Although this never happened, tragedy struck nonetheless. On Christmas Eve 1942, Laddie was suddenly taken ill with bowel cancer and died five days later. He was only fifty-four years old. Just ten days before, he had written a letter to Bill, full of repressed love and emotion, pretending that life at Cavan was going on as normal without him. He spoke of his sore hands full of thistles, the crops being 'a real wash out' and the delivery of a Southdown stud ram for the fat lamb enterprise. Unable to bring himself to ask for details about what Bill was going through, he finished his letter with stifled feelings, 'Very best of luck my boy and love from us all.'

Throughout his adult life, Laddie had been devoted to the management and development of Cavan, keeping in touch with modern methods and possessing an almost boundless energy and enthusiasm for improving the property and continuing the work of his father and grandfather. Apparently, he did not take any prominent part in public affairs, but like many quiet, hard-working men, helped others in a private and unostentatious manner.[24] Bill, his principal heir, was only twenty years old when Laddie died, inheriting the lion's share of the extensive Castle-Roche assets, which remained in trust until Bill was twenty-four. At this point, according to a schedule of assets in Laddie's probate file, Cavan was 14,300 acres, carrying 11,276 sheep, 179 Aberdeen Angus stud cattle (only half of which were owned by Laddie), 82 herd cattle and 25 horses. These assets together with Laddie's half share of Eumbra, which he owned in partnership with his wife's brother-in-law, Thomas Littlejohn, some personal effects and life policies amounted to a dutiable estate of £173,158.

Perhaps the most interesting aspect of Laddie's probate file is that there is no mention of any significant mortgages or borrowings over Cavan at the time of his death, in spite of the enormous damage sustained in the fire of 1939. But there were death duties payable, leading to disagreement as to how the tax should be discharged, especially as there was less than £4,000 cash in the estate. It was at this point that a mortgage was taken out by the executors for about £40,000. Several provisions of Laddie's will apparently also caused upset within the family. Enid had been left a very modest annuity of £600 and Lorna some life insurance policies, but they had also inherited a quarter share each in Eumbra and half each of Laddie's fifty percent interest in the Aberdeen Angus stud. It was the stud in particular that caused a great deal of friction with the executors and within the family, even though it was entered in the probate documents as being worth only £1,885, half of which belonged to Wilfrid Wade. The problem seems to have been that as Bill inherited all the land at Cavan, the Angus stud effectively had no home. Although Enid offered to purchase a parcel of land at Cavan for the stud, this was refused by the executors.

According to Laddie's will, Enid's and Lorna's shares in the stud were to be held in trust and either the stock sold or the business continued by the trustees on their behalf, in conjunction with Wilf or having purchased Wilf's share. It seems, however, that the trustees wanted to sell the stud to help pay death duties as they had tried unsuccessfully to sell any land at Cavan to raise the necessary cash. But as Wilf was missing in action, Enid steadfastly refused to allow the trustees to interfere in any way with his interest or to sell any of the cattle. She was so concerned about the future of the stud and also Lorna's position

— *Lorna Castle-Roche with one of her stud cattle, c. 1945*

generally under Laddie's will, that she wrote several letters to Bill immediately after Laddie's death, asking him to intervene with the trustees.

Whilst Bill was known to be generous to a fault,[25] he was somehow never able to make things sufficiently right with Lorna, who remained disappointed by the way she had been treated. She and her mother left Cavan in 1944, taking the Angus cattle (then 300 head) with them to a new 1,000-acre property, Tottenham, just west of Sydney.[26] This was a brave move, especially as Wilf had still not returned and was now being held as a prisoner of war in Singapore. When he did come back after being discharged from the army on 18 January 1946, he married Enid the same year and they ran the stud together whilst Lorna returned to Yass where she married Robert Phillips, a stock agent with Winchcombe Carson.

BILL CASTLE-ROCHE RETURNS FROM WAR

It took some months for Bill to obtain a discharge from the army on compassionate grounds, which he finally managed to do on 13 April 1943. He had to build a proper case that he was more useful for Australia back on the land, managing Cavan and Eumbra, than remaining in the armed forces. He had been 'a damned fine soldier' according to one of his compatriots,[27] and was promoted in the first few weeks of being in the army. Nonetheless, his war years deeply affected him. He drank heavily for the rest of his life, damaging his health and shortening his life span.

On returning to Cavan shortly after his discharge, Bill was young and inexperienced for the challenges that now faced him. Fortunately, his capable mother, Enid, had been managing the place since her husband's death and the executors had had the good sense to appoint Jim Northey as overseer on whom Bill was able to rely and who remained at Cavan until 1961. Jim's brother-in-law, Ken Saunders, also joined the workforce at Cavan in February 1946, ultimately becoming manager

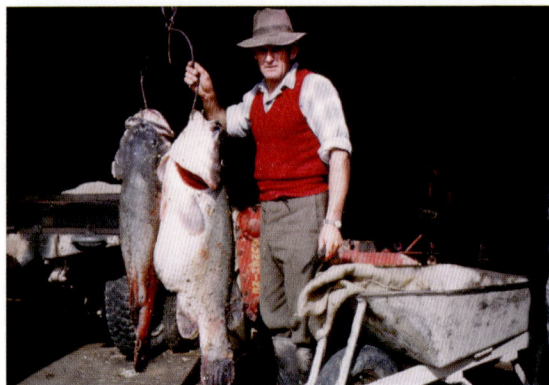

— *Ken Saunders, c. 1950*

and living at Cavan West until he died in 1989.[28] Ken could and would do everything, quickly becoming the 'go to' person on the property. He was a wonderful and talented gardener, growing every manner of vegetable including artichokes, carrots, parsnips, turnips, cauliflower, asparagus and tomatoes, as well as pumpkins which he planted in the flower beds. He was also an apiarist, keeping bees at the back of the cottage on the right-hand side of the main driveway where he lived with his wife and two daughters. He was in charge of fattening the pigs, milking the cows and killing sheep for farm rations; and as a trained diesel mechanic in the Australian air force, he was also responsible for all machinery at Cavan, including the six-ton 'Federal', a truck with a huge water tank on the top for fighting bushfires. In his spare time, Ken was a keen amateur photographer and fisherman, regularly catching huge Murray cod in the Murrumbidgee. In time, Bill came to rely heavily on Ken, who loved Cavan like it was his own and always treated Bill with kindness and the utmost respect.

MARRIAGE

Bill was a strikingly handsome man with an enviable physique, a keen, alert mind and an analytical brain.[29] He was powerful and strong, good enough to become an Olympic swimmer if this had ever been encouraged. Soon after his return to Cavan, he met and married the pretty and vivacious Helen Stewart, the daughter of the chief pastoral inspector and stock buyer for A.B. Triggs in Yass, a successful landowner with properties in New South Wales, Victoria and Queensland. The marriage took place on 14 January 1946 at the Presbyterian church in Yass, but no account of the wedding nor any photographs have

Every anniversary, birthday, Christmas and sporting occasion was an excuse for a party with music, dancing and, of course, a great deal of alcohol. For many of these bacchanalian occasions, guests came from miles around and even drove down from Sydney.

—

survived. The couple had two daughters, Susan ('Susie') born on 18 July 1947 and Gai born on 7 November 1950. Both girls were educated locally and then at Ascham in Sydney.

Helen was several years older than Bill, but more than matched him in terms of her athleticism and zest for life. Like Bill, who was a very good rider, she was a mad keen horsewoman, constantly travelling about the country for shows and equestrian events. She would ride all the way to Gunning, Burrowa and Jerriwa to enter her mounts in hack classes and flag racing, always cutting a most elegant figure. Sharing most of her husband's many sporting pursuits, Helen also played polocrosse with him. Six fields were prepared at Cavan just by the main gate, for regular polocrosse games and for an annual tournament that became a legendary social occasion. Showing a sense of humour, Bill and Helen had a special tournament brooch made in the shape of a worm, which was given to anyone who fell off their horse. After a full day of matches, everyone would gather at the homestead and a rowdy party, fuelled by plenty of alcohol, would ensue well into the night. Not only did polocrosse provide an excuse for such revelry, but tennis matches did too, as Helen was a keen player.

In fact, during Bill and Helen's stewardship, Cavan became a regular playground of parties and social activity, a decadent scene that had not been witnessed at Cavan before. Bill liked to live life to the full after his bitter war experiences, and his wife was only too happy to comply. She was spoilt and flirtatious and he had to stand up to her regularly, but they made a fun-loving and attractive couple. Every anniversary, birthday, Christmas and sporting occasion was an excuse for a party with music, dancing and, of course, a great deal of alcohol. Bill's daughter, Susie, recalls a special drink called 'Kick a Pooh Pooh Punch' that was very popular and made with a bottle each of whiskey, brandy, gin and rum, and sometimes mixed with a tin of peaches. Other famous Cavan brews were 'Happiness Comes Quickly', based on dry martinis and 'Forty Fox Power Juice', a rum-based concoction with brown sugar and raisins that was the favourite tipple on kangaroo shoots.

For many of these bacchanalian occasions, guests came from miles around and even drove down from Sydney. The women dressed up in cocktail gowns and white gloves, purchased from the smartest shops in Double Bay, and the men would wear black tie. Behaviour was often indecorous; on one memorable night, a guest tried crawling through the cat flap in the back door, bringing the whole doorway down as he tried to stand up. No one can remember what food was served at these parties, but it was tradition to bring a dish. Everyone congregated around the enormous bar in the drawing room and eating was rather beside the point! Susie and Gai, both much too young and innocent to be amongst the throng, would watch the noisy and unruly proceedings from the bannisters of the grand staircase, fascinated.

Above — *Sheep at Cavan, c. 1930*
Right — *Cavan shearing shed, c. 1930*

DECLINE

In spite of such excess, Bill was also a serious-minded individual who loved the land and understood the responsibility that his father's untimely death had thrust upon him. Unfortunately, the death duties payable took him all his life to discharge, especially with his and Helen's taste for high living. Working tirelessly every day with a labour force of four or five men, he had a mixed enterprise of merino sheep (with Merryville bloodlines inherited from his father), fat lambs and cattle. According to his remaining diaries which only cover the years 1939, 1946 and 1947, Bill also engaged in a high degree of trading after his father's death, travelling the length and breadth of New South Wales to source almost any breed of cattle and sheep that he felt he could make a turn on. He must have been under enormous pressure, as his diary entry of 18 May 1946 records that he applied for what was in those days a substantial loan of £45,000, which was eventually agreed on 12 August that year.

Like so many pastoralists in that era, Bill also bred racehorses, using a stallion called 'Speargrass', the progeny of 'Spearfelt', a well-known stallion in racing circles. Bill did not have much luck with any winners, but loved racing so much that he carried on breeding racehorses regardless. This was the era of buoyant wool prices and Bill made the most of it. He purchased two Bentleys, one for himself and one for Helen, and drove his carelessly around Cavan as if it was an ordinary work vehicle, dressed in khaki shorts, an open-necked shirt and without a hat. If he found something in the wrong place or an animal that needed attention, whether it was a roll of fencing, a fly-blown sheep or an orphaned calf, he would throw it on the back seat of the car, heedless of the fine leather seats on which his dog, Boogle, also sat.

Bill had little patience for ceremony, being down to earth and egalitarian by nature. He called himself just plain Bill Roche, never using the double-barrelled name of his family. He adored his two daughters and did not allow them to develop airs or graces either. When they went off to school in Sydney, they were sent with the bare minimum of clothing besides their uniform. When at home, he expected them to play and socialise with the children of Cavan employees and never wanted them to be different from their contemporaries. He saw the way the world was going after the war – the breaking down of social class and the rise of egalitarianism. He brought the girls up strictly, to be seen and not heard, but he was always very protective and fair. To a degree, he groomed Susie to be his successor at Cavan, but health was not on his side and, ultimately, he could not keep the property intact for her.

Not ever forgetting the suffering of his fellow countrymen during the war, Bill volunteered to give up 5,000 acres at Cavan for soldier settlement blocks.[30] This was a magnanimous gesture and came at a time when only one estate in the Yass electorate had previously been selected by the government for the scheme. Political expediency had meant that several properties had been purchased or resumed by the Crown in the neighbouring districts of Wagga Wagga and Young, but Yass had been bypassed, in spite of considerable agitation by Bill Sheahan, the popular and outspoken state Labor MP for Yass. Sheahan was a great champion of the soldier settlement scheme, urging the government to accelerate its implementation and calling for land to be acquired in his electorate. When Bill made his land at Cavan available, Sheahan wrote him a personal letter of thanks on 9 May 1946 commending him for his altruism.[31]

Bill was an exceptionally generous man, a great friend to all who knew him and always trying to help others in any way that

Not ever forgetting the suffering of his fellow countrymen during the war, Bill volunteered to give up 5,000 acres at Cavan for soldier settlement blocks.

—

he could. Another example of his benevolent nature was the sale of his block at Mountain Creek in 1956 to a local family in Yass who coveted this particular plot of land with its beautiful flats on the Murrumbidgee and a state-of-the-art shearing shed. Apparently, he sold this to them for a pittance and was justifiably upset when a short time later, they sold the property for a great deal more money.

With these initiatives and other land sales, Cavan was reduced to about 6,000 acres, less than half the land Bill had inherited. He remained plagued by memories of the war and drank to obliterate them. He found sleep difficult and would drink pineapple juice and gin in the middle of the night as a sedative. His health deteriorated and he began to scale down his operations even more, selling Eumbra first in 1963,[32] which he had presumably been managing on behalf of Enid, Lorna and his widowed sister-in-law. Declining further still, Bill realised it was no longer possible for him to carry on at Cavan. Dividing off 3,000 acres of some of the best open and undulating country into Cavan West, he built a single-storey brick house for himself and his family and, in 1966, sold the bulk of Cavan to media giant, Rupert Murdoch, who owns the property to this day.

At Cavan West, Bill soldiered on stoically, finally throwing one last party to celebrate his eldest daughter's engagement. Susie had fallen in love at only twenty-one years old with a

wheat farmer, Peter Scott from Urana near Wagga Wagga, whom she married on 20 July 1968. By this time, her father was in a wheelchair, but he nonetheless attended the wedding. True to character, he wore an ordinary suit instead of a dinner jacket, to avoid standing out and making others, less fortunate than himself, feel uncomfortable. He put on a brave face for the occasion, but must have known how close he was to dying and how the vicissitudes of life had radically altered the future for his daughters. Perhaps with his modern, post-war views, he felt little regret that Cavan was no longer on the scale it had once been and that his girls were lucky enough. Bill survived a few months more; Susie meanwhile announced her pregnancy, but Bill died at the age of only forty-seven on 13 December 1969. It was a tragic end for such a fine man. He was buried in the family vault on the hill above the Cavan homestead, overlooking the home and land that he loved but had been forced to relinquish.

— *Susie Castle-Roche at home at Cavan West*

CHAPTER ELEVEN

THE MURDOCH ERA

A LIFELINE

THERE CAN BE NO DOUBT that Rupert Murdoch's purchase of Cavan was a watershed in its history, a stroke of luck as significant as the moment that William Riley bought it and turned it into a world-class sheep-breeding operation. When so many similar properties had, since the war, fallen into disrepair or disuse or become run down and eaten out through lack of manpower, resources and poor grazing strategies, Cavan's future was secured by the Murdoch acquisition. Here was a man, an Australian, who loved and understood the land, and with the wherewithal and wisdom to bring the property firmly into the twenty-first century – not only adopting new regenerative farming practices and the highest standards of modern pastoralism, but also expanding Cavan to something approaching its original size and thereby achieving economy of scale. Murdoch would not assert that he intended saving Cavan from potential obscurity, acquiring the property chiefly for personal rather than commercial reasons, but with typical Murdoch precision, it has ended up being both his home and a leading wool-growing enterprise, setting new standards in the industry.

OVERVIEW OF THE MODERN ENTERPRISE

Today, Cavan comprises 23,000 acres made up of eight different properties – the original Cavan grant and the neighbouring holdings of Little Plain, Ravenswood, Wyelba, Bloomfield, Saltbox, Cavan West and Woolaroo. Other than Little Plain and Cavan West, none of these properties are situated on the former Cavan run, but lie mostly north or northwest of Cavan

Above — *Rupert Murdoch with his daughter Elizabeth at Cavan, c. 1970*
Previous pages — *Sheep yards at Cavan at sunset*

— *Matthew Crozier, manager of Cavan, with Malcolm Peake, stud manager*

itself. There are eight full-time men on the property, similar to the size of the team in the early days of settlement, including a general manager, Matthew Crozier, and a stud manager, Malcolm Peake.[1] Additional staff are employed in the farm office and the homestead.

Merino sheep and wool production are at the core of the enterprise so that modern Cavan is still true to its original purpose. There are approximately 30,000 sheep in total on Cavan, of which 13,000 are commercial ewes and 2,000 are breeding ewes retained in the stud. Of the commercial ewes, only about 3,000 are joined annually to Poll Dorset rams in a limited, cross-bred, fat lamb operation that is run alongside the much more important wool-growing business. These 'terminal' cross-bred sheep are all sold at around five months, which helps to create extra cash flow and decrease the number of mouths

— *Ram sale at Cavan in 2017*

There are approximately 30,000 sheep in total on Cavan, of which 13,000 are commercial ewes and 2,000 are breeding ewes retained in the stud.

—

that need to be fed over the difficult summer months. The remaining 10,000 commercial ewes are joined to merino rams, their wethers being retained for one or two years before being sold whilst the ewes are classed, with only the best seventy-five percent kept for the commercial flock.

Initially, when Murdoch bought Cavan, the sheep he inherited were based on Merryville bloodlines, themselves descendants of the Riley Saxons. Murdoch then brought in rams from Wanganella at Deniliquin, and joined these to the smaller, but finer-woolled Cavan merinos. The Wanganella flock was based on the Peppin merino from Spain and France, another important strain of merino in Australia and favoured because of its large frame and heavy fleece as well as its ability to thrive in drier inland regions.

Whilst this practice of using Wanganella rams lasted about twenty-five years, a decision was subsequently made in the 1990s to reintroduce sheep with a considerable Saxon influence, purchased from the Bogo stud at nearby Bookham. Behind this change was the fact that the Saxon type is more suited to an area of higher rainfall than the Peppin, which was prone to fleece rot in the wetter climate of the Cavan district. Moreover, it was about this time that the market for fine wool was improving and the peerless quality of wool that continued to be produced by the Saxon merino in terms of fineness, softness and colour, was widely acknowledged as indisputable.

Whilst wool is the focus of the Cavan enterprise, a beef cattle herd comprising about 2,500 head is also maintained. Until about 2005, only Poll Hereford were grazed by the Murdoch family at Cavan, but the beef enterprise was then converted to Angus to take advantage of the market premium for Angus beef prevailing at the time and because it was recognised that this breed handles the Cavan country of hills and limestone rocks more easily than any other. In 2015, Wagyu beef cattle were also introduced with the purchase of ten bulls, which are joined annually to approximately 350 Angus first calving heifers and the progeny sold. As in the case of Angus beef, Wagyu currently commands a premium price, including for cross-bred cattle.

As for cropping, there is very little undertaken at Cavan, not least because the most fertile flats, about 900 acres of loamy, silty, arable land opposite Bloomfield on the river, are often under water because of the operations of the nearby Burrinjuck Dam. Once the dam is about eighty-five percent full, flooding starts to occur, making the sowing of crops a precarious undertaking. When possible, however, oats, wheat, canola and lucerne are planted for grazing stock, but rarely for commercial purposes.

WOOL GROWING AT CAVAN TODAY

Wool growing at Cavan today has reached a peak of excellence and innovation probably only comparable to the Rileys' achievements almost 200 years ago. Whilst there are many similarities with the Riley Saxons in the existing Cavan flock, there are also some important differences. Today's sheep at Cavan are probably three times more productive in terms of weight and quality of fleece than the early Saxons, which since the days of the Riley stud have been highly improved by selective

Top — *Angus stud bull at Cavan*
Bottom — *Cavan-bred Angus heifers in calf to Wagyu bulls*

Today's sheep at Cavan are probably three times more productive in terms of weight and quality of fleece than the early Saxons, which since the days of the Riley stud have been highly improved by selective breeding over many generations.

—

breeding over many generations. Although the fineness of the Cavan wool is probably similar to that of the Riley Saxons, 18 or 18.5 micron, the frame of today's sheep is larger, its fertility has been increased and the wool is longer, stronger and of a greater density.

Most of these changes can be attributed to the introduction at Cavan of the Bogo merino bloodlines and the subsequent purchase of the Bogo stud in May 2014. It is these Bogo genetics, originally based on the Saxon merino, that produce not only fine wool, but a big cut, long staple and fertile animal with a fast-growing fleece. In the case of some sheep, the growth of their wool is so vigorous that it is possible to shear them twice a year. In contrast to the Wanganella sheep, the Bogo animals have been instrumental in reducing the micron of the Cavan wool and improving its colour and lustre.

Sheep breeding at Cavan has become a highly scientific and carefully thought-out process, using every available technology, the latest research and advanced management strategies. The aim is to produce the highest quality and quantity of wool with as few sheep as possible, making both the size of animal and type of fleece important. A moderate-sized frame, a little larger than the original Riley Saxons, is now the goal at Cavan, with

the perfect weight for a mature ewe being between 60 and 65 kilograms. Sheep of this size tend to be more fertile than smaller types and eat less than a larger-framed animal, yet carry enough body fat and muscle reserves to keep reasonably healthy through drier periods in the year. Today's sheep at Cavan are also much less wrinkled or plainer bodied than their original ancestors, an advantage in terms of shearing and keeping flystrike at bay. The traditional theory that more wrinkles lead to more wool growth was proved wrong many years ago.

Another axiom of wool growing that has been disproved since the days of the Rileys is that a small, tight crimp and short staple is an indication of fineness or softness. When micron testing was introduced in the 1960s, there was shown to be little correlation between these factors with the result that at Cavan today, the focus is to produce fleeces with a bigger crimp. Not only does the larger, more loosely waved crimp yield a longer staple, which in turn gives a greater weight of fleece, but the fibre is more robust with a greater tensile strength better suited to modern spinning machines. Density in the fleece or more hair follicles per square centimetre of skin is also important as this, together with staple length, increases the weight of the clip.

Whilst the softness of the wool depends largely on the average diameter or micron of the wool (the lower the micron the softer the wool), the variation of fibre diameter in each fleece is similarly relevant. Every fleece comprises some fibres that are coarser than others and the more compressed this range is or the lower the variation, the softer the wool. As softness is a quality that is universally desired, the Cavan breeding programme aims to produce wool not only with a lower micron count, but also with this narrower variation of fibres to maximise the luxuriousness of its feel and its pliability. Thus, contrary to what the Rileys believed, that only a heavier-framed sheep could produce the finest quality wool combined with the greatest weight of fleece, it is now possible to do this with the smaller, modern merino.

Combining and repeating all these factors in a flock takes time and a strict breeding programme. As in the Riley era, an index is kept at Cavan recording the vital statistics of each sheep including its body weight, fleece weight and micron count. This data, along with regular visual assessments to judge structure, mobility, wool colour and softness, is used to determine which animals to cull and which to keep. Examinations of rams tend to be particularly stringent because they influence the flock to a greater degree than ewes. Through joining, ewes produce only one or two lambs each year, whereas rams impregnate up to sixty ewes per annum, thereby influencing up to 120 lambs by comparison. Additionally, an on-farm sire evaluation programme is run so that the top commercial ewes are artificially inseminated and their progeny run as a peer group to assess whether they are capable of improving the existing flock. Very occasionally, genetics are introduced from outside the stud so that Cavan genetics can be compared to other leading genetics in the industry. The whole operation is state-of-the-art, with the use of genomics on the horizon to assess an animal's potential value and performance.

Shearing takes place in November for the stud flock and in February for the commercial animals, although the wethers are shorn twice a year because the length of their fleece makes it economically viable to do this. There are now four shearing sheds on Cavan, all built by Murdoch, the original sheds being both at Cavan West and on a block at Mountain Creek, which was sold by Murdoch in the early 1970s. In contrast to the methods of the Rileys, the sheep are no longer washed prior to shearing. Instead, the fleece is 'scoured' to remove grease

Above top — *Matthew Crozier sorting wool*
Above — *New Ravenswood shearing shed*

and most of the dirt after clipping. Scouring and subsequent combing of the wool to remove grass seed and vegetable matter is mostly done in China, but also sometimes in the Czech Republic or India.

SHEEP MANAGEMENT

None of the fine wool growing could be achieved at Cavan to such a high standard without proper sheep husbandry. Animals are drenched regularly to prevent worms and moved around into fresh paddocks, but perhaps the most important task undertaken to keep the sheep healthy is the highly controversial practice of mulesing. This is the removal (often without any pain relief) of surplus skin from the breach of the animal around the tail. It was begun in the 1930s as the result of shearers noticing that those sheep that had been accidentally cut in this area were no longer susceptible to flystrike. Blowflies, more specifically *Lucilia cuprina*, are potentially fatal to sheep who, if their fleece is invaded, can contract excruciatingly painful blood poisoning as a result.

Mulesing is only carried out on merino breeds as other sheep types do not have the warm and dirty wrinkled skin that attracts the fly. In the absence of equally good, cost-effective methods to protect sheep, most wool growers in Australia practise mulesing. Since around 2004, however, pressure groups claiming to be concerned with animal welfare have waged aggressive campaigns against the practice, putting significant pressure on the fashion industry to ensure that the procedure is halted. The Australian Wool Board has invested more than AUD50 million in trying to resolve the issue through a number of means including genetics, clipping, chemical solutions and the genetic modification

— *Sheep after mulesing*

of both the fly itself and the various strains of merino. It has been estimated that if mulesing were abandoned tomorrow, Australia would lose 3,500,000 lambs per annum out of a total of 20,000,000 because of flystrike.

At Cavan, the practice of mulesing is on the way out. Through clever and committed breeding, the wrinkle at the sheep's breach is starting to disappear and, already, about twenty-five percent of new-born lambs no longer need the procedure. For the remainder, mulesing (which is now always carried out with pain relief) will cease altogether at Cavan within the next couple of years from the time of writing, a triumph of genetic improvement combined with social awareness. Without question, the steps that have been taken in this area at Cavan are typical of the innovation and fearlessness in adopting modern thinking and strategies to be found throughout the pastoral enterprise.

SUSTAINABILITY AND PASTURE MANAGEMENT

Sustainability is the most important driver as to how wool is grown at Cavan and is at the heart of the grazing and landcare practices utilised on the property. Under the keen eye of Alasdair Macleod, Murdoch's son-in-law, who takes a very personal interest in the running of Cavan and to whom the manager reports, a policy of regenerative landscape management has been implemented which aims to build and maintain healthy, productive soils for the present day and the future. It seeks to restore the land, its nutrients, biodiversity and natural resilience through the adoption of certain strategies, including time-controlled grazing, the reduction of synthetic fertilisers, investment in revegetation and water conservation. With regenerative practices such as these, soil degradation incurred over many decades can ultimately be reversed, increasing the health of the landscape and enabling greater resilience against drought and other adverse weather conditions.

Allan Savory, a Rhodesian ecologist and farmer, was the first to develop this approach to land management. He recognised that droughts were a way of life and that the overstocking and clearing of land, due to the increased density of settlement in rural areas, had led to poor agricultural outcomes. Not everyone in Australia (or in other rural communities) is comfortable with or convinced by these practices, especially as they involve a radical change in organisation and sometimes to infrastructure.

In addition, these changes involve increasing the size of mobs of sheep and moving them more frequently to allow pastures to benefit from longer periods of rest, an alteration in management practices that many find challenging. But plenty of leading graziers have made this shift and are now benefitting from the increased productivity and overall resilience in their pastures that results.

Noticeable changes have already taken place at Cavan with this new policy, especially the use of planned and strategic grazing, whereby large mobs remain for only a limited time in paddocks to avoid the stock grazing selectively on their favourite plants and to encourage them instead to eat all types of grass and herbage. This results in a more even ground cover, leading not only to greater biodiversity and less chance of erosion, but also helping to restore the carbon content in the soil. Carbon is essential for soil fertility, releasing nutrients for plant growth, promoting good soil structure and health, and acting as a buffer against harmful substances. Studies have shown that moderate grazing in drier climates helps to avoid carbon depletion, although the process of sequestering carbon can take decades. In many areas in Australia, soil carbon levels have dropped by up to one fifth of their pre-agricultural levels because of overstocking, cultivation and stubble burning, but even a small improvement in a large area will have a major impact on production capacity.

ECOLOGY

Most of the pastures at Cavan are native and have had only moderate renovation in the past with the planting of some foreign species, including subterranean clover. Nonetheless, previous grazing practices have influenced the species

— *Inside new Cavan shearing shed*

composition of these pastures and their weed burden.[2] To better understand the impact of livestock grazing on the diversity and resilience of pastures over time, a sixty-acre paddock of steep, undulating country is being excluded from use by the Cavan flocks and herds for a number of years, and the changes to the variety and abundance of both native and introduced plants, erosion stabilisation and soil health is being documented.

With livestock excluded from this area, a survey has already been carried out to identify and record all the plant and animal species currently living in the paddock. This baseline information is being monitored each season by an independent ecologist to ensure a full catalogue of existing trees, shrubs, grasses and wildflowers is made, with the emergence of new species that now have the opportunity to germinate also being identified and recorded. Already, remnants of threatened ecological communities have been discovered growing in the exclusion paddock, including box-gum grassy woodland with uncommon understorey species such as chocolate lilies, diggers speedwell, barbed wire grass and Yass daisies. Before white settlement, it is believed that box-gum woodlands were spread widely across the Cavan district and beyond and were dominated by complex alliances of different eucalyptus species of box trees such as yellow, white and grey box, with specific types of gum trees, including Blakely's red gum and river red gum, and that beneath the timber was scattered a wide variety of native grasses, forbs and shrubs. A single mature black cypress pine, now a rarity in the landscape, has also been identified in the Cavan exclusion paddock growing amid the mixed woodland of kurrajongs and other pine species, and there is evidence of recent germination of its seed now that stock has been permanently removed.

Altering the grazing management over this and other

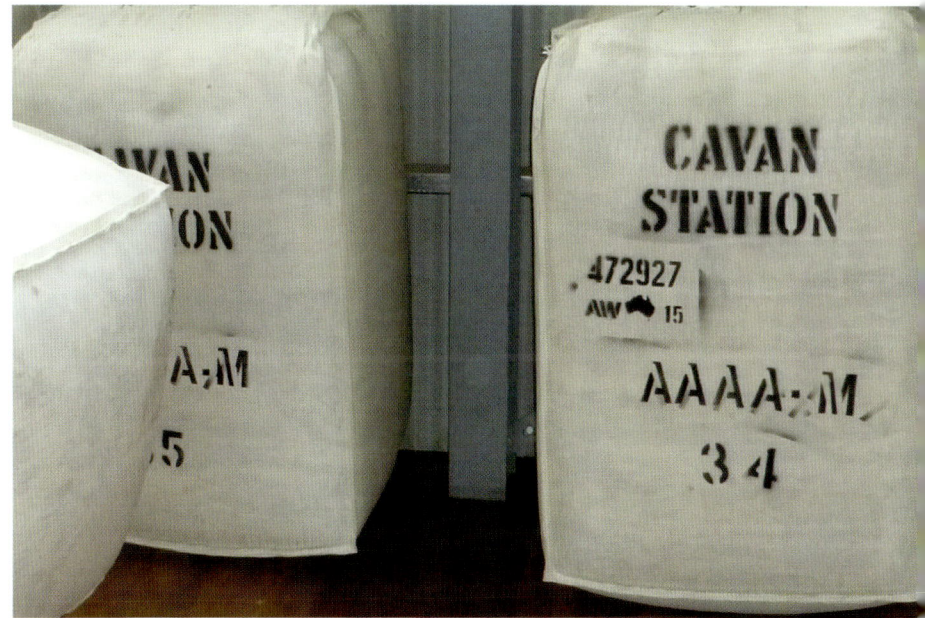

Above top — *Cavan wool bales ready for sale*
Above — *Pasture inspection at Cavan*

paddocks will thus allow for tiny seedlings from threatened species to grow; for a positive, steady change to occur in the balance of pasture grasses over weeds; and for desirable autumn, winter and spring perennial grasses, such as weeping grass, wallaby grass and corkscrew grass, to grow alongside the currently more dominant grasses that flourish in late summer, namely, red leg grass, windmill grass and hairy panic. The latter outcome will prevent a feed gap in winter and early springtime, providing more uniform nutrition for sheep (whose wool fibres can otherwise become uneven in diameter). It will also limit the amount of bare ground, which encourages the growth of weeds, especially saffron thistle, and generally build resilience against drought. As with all these new initiatives at Cavan, there is a palpable excitement amongst the team about the rewards that are being visibly reaped both in the ecologically sound management of the land and in the highly scientific sheep-breeding programme.

THE LONG VIEW

Cavan is now producing a clip of about 800 bales per annum of 190 kilos each and valued at well over AUD1.5 million. It is both a progressive and profitable business and, with its emphasis on sustainability, Cavan wool has no problem finding a buyer. The fashion industry has become very aware of how wool is produced and there is an increasing preference to source product from ecologically sound producers. The Kering Group, owner of well-known luxury brands such as Gucci, Saint Laurent and Stella McCartney, is leading the field in this area, placing sustainability at the core of its business strategy.[3] Not only is Kering aiming by 2025 to source wool, cotton, leather and other

> There is a palpable excitement amongst the team about the rewards that are being visibly reaped both in the ecologically sound management of the land and in the highly scientific sheep-breeding programme.
>
> —

materials exclusively from suppliers whose primary producers are committed to regenerative rather than extractive farming processes, but it has also already established a tangible measure of improvement in environmental outcomes. Having pioneered the methodology of an environmental profit-and-loss account which assesses the impact of its business on, amongst other things, water use, ecosystems and climate, Kering is able to make informed decisions about and steer its operations towards those suppliers and producers that have achieved the highest levels of environmentally friendly performance.

Fashion brands are also extremely conscious of animal welfare and this has become another driver for traceability in their supply chain, sourcing materials from known suppliers whose production methods are ethically sound.[4] In relation to wool production especially, fashion brands and retailers favour wool growers that have discontinued the practice of mulesing, a response to the young generation of millennials who are highly sensitive to this issue. Thus, in so many respects, the wool that is produced at Cavan aligns with the new business philosophy of leading fashion houses, not only because of its outstanding quality, but also due to Cavan's wholehearted commitment to sustainable farming. Moreover, with increasing interest by

The wool that is produced at Cavan aligns with the new business philosophy of leading fashion houses, not only because of its outstanding quality, but also due to Cavan's wholehearted commitment to sustainable farming.

—

the fashion industry in the provenance of wool, the Murdoch family is now exploring ways of marketing their product direct to clothing manufacturers – yet another innovation in its overall sheep enterprise.

Further reason for optimism at Cavan is that wool market conditions are currently extremely buoyant. Wool is once again widely favoured by the fashion industry, partly as a natural, biodegradable material, and also because of its being an environmentally friendly one when produced, as it is at Cavan, by sustainable and regenerative methods. It is also recognised as being comfortable, versatile, breathable and elegant, so that a generation of consumers, brought up on synthetics, has learnt the benefits of merino wool through successful marketing of these features. Moreover, the fibre advocacy campaign led by the Woolmark Company has helped to make wool the product of choice in the rapidly growing Chinese market for luxury goods. Whilst the Manton, Riley and Castle-Roche families saw the full cycle of boom and bust over seven generations, in the absence of any prolonged period of drought, Cavan's wool-growing operation looks set to enjoy a more stable and rosy future.

At least ninety percent of the world's fine apparel wool is grown in Australia and there are no serious competitors on the horizon. Amounting to approximately 340 million kilograms each year, this quantity cannot possibly be rivalled by any other country in the foreseeable future. Most of the wool grown in Australia is exported and is worth about AUD3.6 billion annually. Just a few years ago, the same quantity of wool would have been worth only a little more than half this figure, an indication of the enormous growth in the price of wool in the past eight years and of new markets opening up that historically used only synthetic materials. One such market is sports clothing, which increasingly uses wool for layering products, but there is also the emerging 'athleisure wear', a style of clothing that can either be worn for exercise or casually.

Taking all this into account, the view over Cavan itself and what became popularly known as Riley's run is as lofty as the wool market is today. Looking down from Murdoch's favourite spot at Ravenswood, over the original Cavan grant across the Murrumbidgee, the property appears in outline much as it would have done when Hume and Hovell crossed it in 1824. Extending uninterrupted from the river bank south towards the horizon, a mantle of grass spread over its contours, the Cavan landscape is not only wild and beautiful, but most importantly, is full of promise. Once described by Thomas Mitchell[5] as 'beyond description lovely', it continues to arrest the onlooker, especially in spring with its 'flowery meadows', 'umbrageous woods' and limestone outcrops. Thriving under the auspices of the Murdoch family, whose intent it is to hand Cavan down in the best of health to the next generation, the future seems as bright as it was to the Rileys almost two centuries ago.

At least ninety percent of the world's fine apparel wool is grown in Australia and there are no serious competitors on the horizon. Amounting to approximately 340 million kilograms each year, this quantity cannot possibly be rivalled by any other country in the foreseeable future.

—

— View over the Murrumbidgee River at Cavan, looking west

ENDNOTES

INTRODUCTION

1. Neil Chenoweth, *Rupert Murdoch, the untold story of the world's greatest media wizard*, 2001, p 146.

CHAPTER ONE

1. I am indebted to Dr Gavin Young, Honorary Visiting Research Fellow of The Australian National University, Canberra, for most of the information on geology and palaeontology in this chapter.
2. *The Sydney Morning Herald*, 11 and 24 May 1870.
3. *Quarterly Journal Geological Society*, Vol. 4, 1848, pp. 63–66.
4. Thomas L. Mitchell, *Three expeditions into the interior of eastern Australia, with descriptions of the recently explored regions of Australia Felix, and of the present colony of New South Wales*, 1838.
5. Robert Etheridge, 'Report on the limestone caves at Cave Flat, junction of the Murrumbidgee and Goodradigbee Rivers, County of Harden', 1889, p 36.
6. For an excellent account of Shearsby's life, see: P.R. Bindon & A.J. MacQuillan, *Life and Times of A.J. Shearsby, Yass, 1872–1962*, 2016.
7. Ibid, p 45.
8. For a detailed discussion on the Cavan limestone see: P.J. Koluzs, 'Sedimentology and environmental reconstruction of the Cavan limestone, Taemas district, N.S.W.', unpublished thesis, The Australian National University, Canberra, 1972.
9. Gavin C. Young, 'Placoderms (armoured fish): dominant vertebrates of the Devonian Period', *Annual Review of Earth and Planetary Sciences*, 2010, pp 523–550.
10. Gavin C. Young, 'Wee Jasper–Lake Burrinjuck fossil fish sites: scientific background to National Heritage Nomination', Proceedings of the Linnean Society of New South Wales, 2011.
11. Ibid, p 94.
12. George Bennett, *Wanderings in New South Wales, Batavia, Pedir Coast, Singapore and China; being the journal of a naturalist in those countries, during 1832, 1833 and 1834*, 1834, p 189.
13. Thomas Walker, *A month in the bush of Australia: journal of one of a party of gentlemen who recently travelled from Sydney to Port Philip with some remarks on the present state of the farming establishments and society in the settled parts of the Argyle Country*, 1838 (J. Cross, London), pp 12–13.
14. Information provided by the Australian Institute of Aboriginal and Torres Strait Islander Studies.
15. For a detailed account of the Ngunawal activities in the Canberra region, see Lyall Gillespie, *Aborigines of the Canberra region*, 1984.
16. Thomas L. Mitchell, *Three expeditions into the interior of eastern Australia: with descriptions of the recently explored region of Australia Felix and of the present colony of New South Wales*, 1838, p 340.
17. Heather Goodall, 'A history of Aboriginal Communities in N.S.W. 1909–1939', unpublished thesis, University of Sydney, 1982.
18. Peter R. Kabaila, *Wiradjuri Places, The Murrumbidgee River Basin*, 1995.
19. W.K. Hancock, *Discovering Monaro*, 1972, p 57.

20. All information on ecology provided by Alison Elvin of Natural Capital Pty Ltd.

21. Henry W. Haygarth, *Recollections of a Bush Life in Australia during a Residence of Eight Years in the Interior, 1848*, pp 120–121.

CHAPTER TWO

1. Thomas M. Perry, *Australia's first frontier: the spread of settlement in New South Wales, 1788–1829, 1963*, pp 28–33.

2. Reproduced in *Hume and Hovell 1824*, edited by Alan E. J. Andrews, 1981, p 77.

3. Ibid, p 299.

4. Ibid, p 288.

5. Ibid, p 287.

6. Ibid, p 79.

7. Ibid, p 289.

8. Ibid, p 299.

9. James Atkinson, *An Account of the State of Agriculture and Grazing in New South Wales, 1826*, p 138.

10. Ibid, p 6.

11. Ibid, p 20.

12. Thomas M. Perry, *Australia's first frontier: the spread of settlement in New South Wales 1788–1829, 1963*, p 36.

13. The original grant is missing from the official records, but is nonetheless evidenced by an entry in the Grants Register kept by the Colonial Secretary, Register No. 71 Folio 261. It is also recorded in a volume entitled the Return of all Grants or Appropriation of Crown Lands without purchase 19 Dec 1825 – 31 Dec 1837.

14. Robert H. Webster, *Currency Lad: The Story of Hamilton Hume and the Explorers*, 1982, p 56.

15. *Hume and Hovell 1824*, edited by Alan J. E. Andrews, 1981, p 247.

16. In the possession of Susie Castle-Roche, great-great granddaughter of Joseph Castle.

17. Depasturing licence number 85 dated 23 April 1838.

18. Kenneth Loiterton, *Cootamundra History*, 2011.

19. The *Sydney Herald*, 6 May 1841.

20. For a detailed history of the Manton family see: W. Keith Neal and D.H.L. Black, *The Mantons: Gunmakers*, 1967.

21. Ibid, p 24.

22. Ibid, p 212.

23. Ibid, p 214.

24. *The Argus*, 27 February, 2, 16, 20 and 27 March 1849.

25. Jill Ker, 'The Wool Industry in New South Wales 1803–1830', *Bulletin of the Business Archives Council of Australia*, 1961.

26. *The Cambridge Economic History of Australia*, edited by Simon Ville and Glenn Withers, 2015, p 162.

27. Stephen H. Roberts, *The Squatting Age in Australia 1835–1847*, 1964, p 47.

28. Ibid, p 46.

29. Information from Marg Kaan, descendant of the Manton family.

30. George Bennett, *Wanderings in New South Wales, Batavia, Pedir Coast, Singapore and China being the journal of a naturalist in these countries during 1832, 1833 and 1834*, 1834, p 168.

31. Ibid, p 168.

32. Eastern Division Cavan Occupation Licence and Pastoral Holding files, number 364. See also: Runs Descriptions in *New South Wales Government Gazette*, 30 September 1848; and Lockyer Potter, 'Cavan', *Journal of the Royal Australian Historical Society*, Vol. 26, 1941, p 188.

33. In the possession of Susie Castle-Roche.

34. Depasturing licence number 415, 26 September 1844 and 25 September 1845. Beckham claimed to have held the licence since 1843, but the licence itself is missing from the official records.

CHAPTER 3

1. Letter dated 1 November 1827, Riley papers, Mitchell Library, Sydney.

2. This is relied on by Charles H. Timperley, *Dictionary of Printers and Printing*, 1839, p 908, but Timperley is not always accurate.

3. Jacob Simon, *British Artists' Suppliers 1650–1950*, National Portrait Gallery, London, 2011.

4. Ian Maxted, *The London Book Trades 1775–1800*, 1977.

5. Ibid, p viii.

6. See generally: Hannah Barker, *Newspapers, Politics and Public Opinion in late Eighteenth Century England*, 1998.

7. A. Aspinall, *Politics and the Press 1780–1850*, 1949, p 41.

8. Ian Maxted, *The London Book Trades 1775–1800*, 1977, p viii.

9. Frank A. Mumby, *Publishing and bookselling from the earliest times to 1870*, 1974, p 185.

10. *St James's Chronicle*, 3 January 1775.

11. *The Times,* 11 April 1788.

12. Charles H. Timperley, *A Dictionary of Printers and Printing*, 1839, p 908.

13. For details of Fenn Kemp's career and Alexander Riley's early life in Van Diemen's Land, see: Nicholas Shakespeare, *In Tasmania*, 2004.

14. Jill Ker, 'Merchants and Merinos', *Journal of the Royal Australian Historical Society*, Vol. 46, 1960, pp 206–221. Most of the subsequent details about Alexander Riley's activities and the wool industry are taken from this article.

15. J.C. Garran, *Merinos, Myths and Macarthurs*, 1985, p 132.

16. Letter Alexander Riley to Edward Riley, 15 June 1825, Riley papers, Mitchell Library, Sydney.

17. J. C. Garran, *Merinos, Myths and Macarthurs*, 1985, p 143.

18. Jill Ker, 'Merchant and Merinos', *Journal of the Royal Australian Historical Society*, Vol. 46, 1960, p 217.

19. For full details of Dutton's early life, see: J.C. Garran, *Merinos, Myths and Macarthurs*, 1985, pp 150–158.

20. Ibid, p 132.

21. Ibid, p 153.

22. Ibid, p 162.

23. Ibid, p 132.

24. Letter Alexander Riley to Edward Riley dated 2 September 1827, Riley papers, Mitchell Library, Sydney.

25. Letter Alexander Riley to William Riley dated 24 July 1833, Riley papers, Mitchell Library, Sydney.

26. Charles Massy, *The Australian Merino*, 1990, p 34.

27. Jill Ker, 'Merchants and Merinos', *Journal of the Royal Australian Historical Society*, Vol. 46, 1960, p 217.

28. Anonymous, 'Description of merino sheep and wool 1828-1832' in the Mitchell Library, Sydney, ML B449-451.

29. Charles Massy is of a different view, see: *The Australian Merino*, 1990, p 89.

30. *The Australasian*, 17 June 1865.

31. Keith R. Binney, *Horsemen of the First Frontier 1788-1900*, 2005, p 102.

32. Charles Massy, *The Australian Merino*, 1990, p 88.

CHAPTER 4

1. Riley papers, Mitchell Library, Sydney.

2. Letter the Colonial Secretary to Alexander Riley 14 July 1825.

3. In the Mitchell Library, Sydney and also helpfully reproduced in part by James Jervis in *The Journals of William Edward Riley, Journal of the Royal Australian Historical Society*, Vol. 32, 1946, pp 217–268.

4. Letter William Dutton to Governor Bourke dated 6 February 1832.

5. Charles Massy, *The Australian Merino*, 1990, p 88.

6. Ibid, p 89.

7. Ibid, p 88.

8. See generally: Riley papers, Mitchell Library, Sydney.

9. J.C. Garran, *Merinos, Myths and Macarthurs*, 1985, p 157.

10. Also spelt 'Cooradigby'.

11. Grant 151, Register 71, Folio 15, dated 31 May 1841, in the Register of Land Grants and Leases 1792–1856.

12. George Bennett, *Wanderings in New South Wales, Batavia, Pedir Coast, Singapore and China being the journal of a naturalist in those countries during 1832, 1833 and 1834*, 1834.

13. Ibid, p 344.

14. Ibid, p 346.

15. Riley papers, Mitchell Library, Sydney.

16. *The Sydney Gazette* and *New South Wales Advertiser*, 5 July 1834.

17. *Journal of the Yass Historical Society*, Vol. 27, No 3, September 2011.

18. Stephen H. Roberts, *The Squatting Age in Australia 1835–1847*, 1964, pp 56–57.

19. J. C. Garran, *Merinos, Myths and Macarthurs*, 1985, p 155.

20. *The Sydney Morning Herald*, 5 September 1846.

21. Riley papers, Mitchell Library, Sydney.

22. J.C. Garran, *Merinos, Myths and Macarthurs*, 1985, p 144.

23. James Atkinson, *An Account of the State of Agriculture and Grazing in New South Wales*, 1826, p 73.

24. *The Australasian Town and Country Journal*, 29 December 1894.

25. *Sydney Gazette*, 10 December 1836.

26. Probate files of William Riley.

27. Affidavit of Mary Wilson dated 30 March 1837 in the probate files of William Riley.

28. Charles Massy, *The Australian Merino*, 1990, pp 82–83.

29. They were subsequently advertised for sale by Lockyer in the *Sydney Herald*, 21 January 1841.

30. *Maitland Mercury*, 31 January 1855 and *Tenterfield Empire*, 4 February 1856.

31. *A month in the bush of Australia: journal of one of a party of gentlemen who recently travelled from Sydney to Port Philip: with some remarks on the present state of the farming establishments and society in the settled parts of the Argyle country*, 1838.

32. *Commercial Journal and Advertiser*, 13 April 1839.

CHAPTER FIVE

1. Probate files of Honoria Riley.
2. Probate files of William Riley, Affidavit of Stuart Donaldson, dated 15 June 1839.
3. Lockyer Potter, 'Cavan', *Journal of the Royal Australian Historical Society*, Vol. 26, 1941, p 189.
4. James Jervis, 'Settlement in the Marulan-Bungonian District', *Journal of the Royal Australian Historical Society*, Vol. 32, 1946, pp 129–130.
5. James Jervis, 'The Journals of William Edward Riley', *Journal of the Royal Australian Historical Society*, Vol. 32, 1946, p 223.
6. Colonial Secretary's file, Riley papers, Mitchell Library, Sydney.
7. Depasturing licences 415 and 450, dated 26 September 1844 and 25 September 1845 respectively.
8. See generally for this dispute: Colonial Secretary's file, Riley papers, Mitchell Library, Sydney.
9. Letter by Bingham to the Colonial Secretary, dated 1 May 1844.
10. Letter by Bingham to the Colonial Secretary, dated 11 May 1844.
11. Beckham v. Potter, reported in *The Sydney Morning Herald*, 16 February 1848.
12. Edmund Lockyer, *Common Place Book*, Mitchell Library, Sydney.
13. Shirley Finnel, *The Potters of Cavan*, 2001, available at the Yass Archives.
14. *The Australian*, 20 July 1839.
15. Shirley Finnel, *The Potters of Cavan*, 2001, available at the Yass Archives.
16. James Mullyon, 'Henry O'Brien of Douro', *Yass Tribune*, 2 February 1939.
17. Passenger list of the *William Nicol*, 1837.
18. *Yass Evening Tribune*, 17 November 1927.
19. *Yass Evening Tribune*, 4 December 1919.
20. Ibid.
21. Jill Roe, *Stella Miles Franklin*, 2008, p 17.
22. Obituary of Emily Potter, *Tumut Advocate*, 20 March 1906.
23. Ruling, dated 31 May 1855 of W. M. Manning in the Intestacy files of George Potter.
24. See generally: George Thomas Potter, Intestate Case Papers.
25. *The Sydney Morning Herald*, 12 September 1853.
26. Probate files of William Riley.
27. Report by the Colonial Land and Emigration Office, dated 9 October 1851, Colonial Secretary's file, Riley papers, Mitchell Library, Sydney.
28. Terry Kass, 'History of Raby', report written 2002, p 12, in the Mitchell Library, Sydney.
29. Letter from Lord Grey to Governor Darling, dated 14 September 1850, Riley papers, Mitchell Library, Sydney.
30. *Sydney Gazette*, 27 February 1840.
31. *Bell's Life*, 1 January 1859.
32. Terry Kass, 'History of Raby', report written 2002, p 13, in the Mitchell Library, Sydney.
33. New South Wales Land Registry, mortgage numbers 685 and 686, Book 74, dated 30 June 1858.
34. Information on the first 24 students, University of Sydney archives.
35. *The Sydney Morning Herald*, 30 May 1855.
36. *The Sydney Morning Herald*, 17 January 1862.
37. Terry Kass, 'History of Raby', report written 2002, p 13, in the Mitchell Library, Sydney.

38. Ibid, p 14.

39. Letter by Alexander Raby Riley to the Colonial Secretary, dated 31 March 1855, and see generally: Colonial Secretary's file, Riley papers, Mitchell Library, Sydney.

40. *New South Wales Government Gazette*, 13 June 1865.

41. Probate file of Alexander Raby Riley.

CHAPTER SIX

1. Old System Vendors' Index 1825–1984, Book 79 no. 270.

2. Deposit 7, Cavan Station, Box 16, Noel Butlin Archives, Canberra.

3. Ibid.

4. *The Sydney Morning Herald*, 8 March 1862, *Empire,* 27 June 1862, and *Goulburn Herald*, 2 July 1862.

5. For an excellent account of Calvert's contribution to the Leichhardt expedition see: Patricia Clarke, *Pioneer Writer: The life of Louisa Atkinson: novelist, journalist, naturalist,* 1990, pp 168–175.

6. Ludwig Leichhardt, *Journal of an overland expedition in Australia, from Moreton Bay to Port Essington: a distance of upwards of 3000 miles during the years 1844–1845,* 1847, p 135.

7. Ibid, pp 23, 54, 57, 80.

8. Deposit 7, Cavan Station, Boxes 2 and 3, Noel Butlin Archives, Canberra.

9. *New South Wales Government Gazette*, 14 June 1859.

10. Deposit 7, Cavan Station, Boxes 2 and 3, Noel Butlin Archives, Canberra.

11. Obituary of James Calvert, *Australian Town and Country Journal*, 26 July 1884.

12. *Sydney Mail*, 1 April 1871.

13. Portion 12, Parish of Cavan Map Second Edition 1896.

14. Sands Postal Directory.

15. For an excellent biography see: Patricia Clarke, *Pioneer Writer: The life of Louisa Atkinson: novelist, journalist, naturalist*, 1990. See also: Janet Cosh papers, National Library of Australia, Canberra.

16. James Calvert gave his usual address as Cavan on his marriage certificate.

17. See generally: Patricia Clarke, *Pioneer Writer: The life of Louisa Atkinson: novelist, journalist, naturalist*, 1990.

18. J.C. Garran, *Merinos, Myths and Macarthurs*, 1985, pp 133–135.

19. Charlotte Atkinson was the author of the first book for children written in Australia, *A Mother's Offering to her Children.*

20. Patricia Clarke, *Pioneer Writer: The life of Louisa Atkinson: novelist, journalist, naturalist*, 1990, p 3.

21. Rev. William Woolls, 'A sermon preached at St Peter's Church, Richmond, in memory of the late Mrs Calvert 12 April 1874', p 8.

22. Ibid, p 9.

23. Louisa Atkinson, *Gertrude the Emigrant*, edited by Elizabeth Lawson, 1998, p xviii.

24. Susanna De Vries-Evans, *Pioneer Women*, 1987, p 229.

25. Patricia Clarke, *Pioneer Writer: The life of Louisa Atkinson: novelist, journalist, naturalist*, 1990, pp 133–140.

26. Susanna De Vries-Evans, *Pioneer Women*, 1987, p 229.

27. Dated 11 November 1858 and calling for the continuation of the search for the explorer and his party in Port Essington.

28. Letter James Calvert to the Editor, *The Sydney Morning Herald*, 11 November 1858.

29. J.H. Maiden, *Records of Australian Botanists*, 1909, p 83.

30. Louisa Atkinson, *Gertrude the Emigrant*, edited by Elizabeth Lawson, 1998.

31. Rev. William Woolls, 'A sermon preached at St Peter's Church, Richmond, in memory of the late Mrs Calvert 12 April 1874', p 15.

32. Calvert papers, National Library of Australia, Canberra.

33. Atkinson papers, Mitchell Library, Sydney.

34. Deposit 7, Cavan Station, Box 16, Noel Butlin Archives, Canberra.

35. Ibid.

36. Gloria Carlos, 'Taemas – a tale of two bridges', *Boongaroon*, Vol. 30/No. 3, 2014, Yass Archives.

37. Janet Cosh papers, National Library of Australia, Canberra.

38. Obituary of Mrs J.S. Calvert, *Australian Town and Country Journal* 30 November 1878.

CHAPTER SEVEN

1. Deposit 7, Cavan Station, Boxes 2–3, Noel Butlin Archives, Canberra.

2. See generally: Sheila A. Mason, *Nottingham Lace 1760s–1950s*, 1994.

3. The *Nottingham Directory: being a general list of the merchants, tradesmen and principal inhabitants of Nottingham, New Radford and New Snenton*, 1815, p 84. Also, Glover, *Nottingham Directory*, 1825, p 100.

4. Will of Joseph Castle 9 March 1831 in the Probate file of Joseph Frederick Castle.

5. *Perry's Bankrupt and Insolvent Gazette*, 30 December 1837.

6. Diary of Joseph Castle, 1838–1852, Deposit 7, Cavan Station, Box 31, Noel Butlin Archives, Canberra.

7. *The Sydney Morning Herald*, 28 December 1846.

8. Deposit 7, Cavan Station, Box 4, Noel Butlin Archives, Canberra.

9. Albert B. Piddington, *Worshipful Masters*, 2000, p 151.

10. Ibid.

11. Ibid, p 152.

12. *The Sydney Morning Herald*, 7 January 1856.

13. *The Sydney Mail*, 20 April 1921.

14. Ibid.

15. Albert B Piddington, *Worshipful Masters*, 2000 p 153.

16. Ibid, p 155.

17. Ibid.

18. Deposit 7, Cavan Station, Boxes 1–3, Noel Butlin Archives, Canberra.

19. Joan Frew, *Queensland Post Offices 1842–1980*, 1981. Also, notes written by Bill Castle-Roche and an unidentified newspaper obituary both in the possession of Susie Castle-Roche.

20. For example, *The Sydney Morning Herald*, 27 April 1857 and 16 May 1857.

21. Deposit 7, Cavan Station, Box 2, Noel Butlin Archives, Canberra.

22. *The Sydney Morning Herald*, 24 December 1870.

23. Deposit 7, Cavan Station, Box 26, Noel Butlin Archives, Canberra.

24. *The Tumut Advocate and Farmers and Settlers' Adviser*, 10 March 1925.

25. Information kindly provided by Judith Kelly, great-great-granddaughter of John Thatcher.

26. Portion 12, Parish of Cavan Map, Second Edition 1896.

27. Janet Cosh papers, National Library of Australia, Canberra.

28. Obituary of John Thatcher, *The Tumut Advocate and*

Farmers and Settlers' Adviser, 10 March 1925.

29. Parish of Cavan Map, Fifth Edition, 1968.

30. Probate file of John Frederick Thatcher, Schedule No. 5, Inventory of Debts due to the Estate.

31. Information kindly provided by Cameron Archer, Thatcher descendant.

32. Portions 82 and 83, Parish of Cavan Map, Second Edition 1896.

CHAPTER EIGHT

1. Governor Lachlan Macquarie, *Journal of the Royal Australian Historical Society*, Vol. 16, 1931, pp 434–435.

2. See generally: Stephen H. Roberts, *The Squatting Age in Australia 1835–1847*, 1964.

3. See generally: Robert Hughes, *The Fatal Shore*, 1987.

4. R.H. Webster, *Currency Lad: The Story of Hamilton Hume and the Explorers*, 2012, p 209.

5. Letter Hamilton Hume to William Riley 18 February 1834, Riley papers, Mitchell Library, Sydney.

6. *Goulburn Herald*, 27 September 1851, 24 April 1852 and 19 April 1856.

7. Riley papers, Mitchell Library, Sydney.

8. The full title of Walker's journal: *A month in the bush of Australia: journal of one of a party of gentlemen who recently travelled from Sydney to Port Philip: with some remarks on the present state of the farming establishments and society in the settled parts of the Argyle country.*

9. *Australian Town and Country Journal*, 26 January 1878.

10. James Mullyon, 'Henry O'Brien of Douro', *Yass Tribune*, 2 February 1939.

11. Ibid.

12. Stephen H. Roberts, *The Squatting Age in Australia 1835–1847*, 1964, pp 212–213.

13. Journal of William Riley from Raby to Yass Plains, 1829–1830, Mitchell Library, Sydney.

14. George Bennett, *Wanderings in New South Wales, Batavia, Pedir Coast, Singapore and China being the journal of a naturalist in these countries during 1832, 1833 and 1834*, 1834, pp 184–186.

15. James Mullyon, 'Henry O'Brien of Douro', *Yass Tribune*, 2 February 1939.

16. Obituary of Rees Jones, *Yass Courier*, 24 August 1886.

17. 'Reminiscences of Ann Jones 1810–1903', p 51, Yass Archives, Yass.

18. For example: depasturing licences 679 and 732 dated 3 May 1843.

19. *Sydney Herald*, 30 June 1842.

20. 'Reminiscences of Ann Jones 1810–1903', pp 56–59, Yass Archives, Yass.

21. For a full account of his contribution to Yass, see *Yass Courier*, 24 April 1861

22. *Old System Vendors' Index Book* 79, Number 248.

23. 'Reminiscences of Ann Jones 1810–1903', Yass Archives, Yass.

24. *Yass Courier*, 27 March 1858.

25. P. E. Leroy, Samuel Terry, *Journal of the Royal Australian Historical Society*, Vol. 47, 1961, pp 281–291.

26. Ibid, p 287.

27. Governor Lachlan Macquarie, *Journal of the Royal Australian Historical Society*, Vol. 16, 1931, p 435.

28. Testimony of Riley in 'Report of the Commissioner of Inquiry, on the state of agriculture and trade in the colony of New South Wales', Thomas Bigge, 1823, Appendix K4, p 16.

29. Journal of William Riley from Raby to Yass Plains, 1829–1830, Mitchell Library, Sydney.

30. Beryl Pittman, *From a Little Acorn: A History and Genealogy of the Davis Family of Gounyan*, 1989, p 42. Also Portion 228, Parish of Warroo Map, Second Edition, 1904.

31. Memoirs of Elizabeth Barrett Davis transcribed in Beryl Pittman's *From a Little Acorn: A History and Genealogy of the Davis family of Gounyan*, 1989, pp 324–343.

32. Old System Vendors' *Index Book* 67 Number 165.

33. James Mullyon, *Yass Tribune*, 1939.

34. *Yass Courier*, 20 March 1859.

35. *New South Wales Government Gazette*, 30 September 1848.

36. Information from the Yass Historical Society.

37. Cavan Public School file,1877–1888, pp 4–5.

38. Ibid, p 20.

39. Ibid, p 6.

40. Ibid, p 22.

41. Ibid, p 34.

42. Ibid, p 32.

43. Portion 57, Parish of Taemas Map, First Edition, 1884.

44. Cavan Post Office file 1876–1916.

45. Ibid, p 59.

46. Ibid, p 65.

47. All information about the development of Yass from the Yass Archives, Yass.

48. Reverend Brian Maher, *Memories of Yass Mission, St Augustine's Parish, Yass, 1838-1988*, 1988, p 14.

CHAPTER NINE

1. Obituaries of Joseph Castle, *Southern Argus*, 24 May 1883 and *The Gundagai Times*, 29 May 1883.

2. Will of Joseph Frederick Castle dated 4 July 1874.

3. Deposit 7, Cavan Station, Box 26, Noel Butlin Archives, Canberra.

4. In the Yass Archives.

5. Deposit 7, Cavan Station, Box 2, Noel Butlin Archives, Canberra.

6. The Cavan diaries mention his attending Windeyer's funeral in December 1874.

7. Deposit 7, Cavan Station, Box 7, Noel Butlin Archives, Canberra.

8. Ibid.

9. Portion 135 Parish of Taemas Map, first edition, 1884.

10. A beautiful property comprising 1,780 acres of freehold and a leasehold area with ten miles of river frontage on the Murrumbidgee. See: *The Sydney Morning Herald*, 1 February 1860.

11. Deposit 7, Cavan Station, Box 7, Noel Butlin Archives, Canberra.

12. Deposit 7, Cavan Station, Box 2, Noel Butlin Archives, Canberra.

13. Deposit 7, Cavan Station, Box 7, Noel Butlin Archives, Canberra.

14. Deposit 7, Cavan Station, Box 7, Noel Butlin Archives, Canberra.

15. Agreement between F.W.C. Roche and John Styles dated 31 March 1899, Deposit 7, Cavan Station, Box 16, Noel Butlin Archives, Canberra.

16. Information kindly provided by Judith Sime, Styles descendant.

17. Only a very short obituary appeared in the *Yass Courier*, 8 August 1899.

18. Probate file of Wilhelmina Castle.

19. Obituary, Joseph Alexander Kethel, *Journal of the Royal Australian Historical Society*, Vol. 32, p 211.

20. Deposit 7, Cavan Station, Box 7, Noel Butlin Archives, Canberra.

21. The site of the original cemetery at Cavan is not known, but was probably near the river by the old homestead.

22. Deposit 7, Cavan Station, Box 5, Noel Butlin Archives, Canberra.

23. Will dated 12 July 1896, Probate file of Eliza Castle Roche.

CHAPTER TEN

1. *Goulburn Evening Penny Post*, 17 March 1906.

2. Sydney University Archives.

3. Information kindly provided by Lorna Phillips, great granddaughter of Joseph Castle.

4. I am indebted to Lorna and Tim Phillips, respectively great granddaughter and great grandson of W.F.F. Castle-Roche, for most of the information about Laddie and his activities.

5. See generally: David Moeller, *The Merryville Type*, 2003.

6. Information kindly provided by Roz Waters, daughter of Jim Northey.

7. *Inverell Times*, 23 January 1925.

8. Information kindly provided by Lorna Phillips.

9. Obituary of Abraham Wade, *Yass Courier*, 20 July 1911.

10. Geo. A. List & Sons, *The Leading Studs of Australia*, 1939, pp 230–232.

11. See generally: Bobbie Hardy, *From the Hawkesbury to the Monaro, The Story of the Badgery Family*, 1989.

12. *Yass Courier*, 28 May 1925.

13. *Wagga Wagga Daily Advertiser*, 28 May 1925.

14. *Yass Courier*, 1 June 1925.

15. Gloria Carlos, 'Taemas – a tale of two bridges', *Boongaroon*, Vol. 30, Number 3, September 2014.

16. Ibid.

17. In the possession of his daughter, Susie Castle-Roche, also great-great granddaughter of Joseph Castle.

18. *The Sydney Morning Herald*, 17 January 1939 and 21 February 1939.

19. Information kindly provided by Lorna Phillips.

20. In the possession of Susie Castle-Roche.

21. Information kindly provided by Lorna Phillips.

22. In the possession of Susie Castle-Roche.

23. Correspondence in the possession of Susie Castle-Roche.

24. Undated obituaries for Laddie Castle-Roche in the possession of Susie Castle-Roche.

25. Obituary of Bill Castle-Roche, *Yass Courier*, 16 December 1969.

26. *The Land*, 20 July 1945.

27. Obituary of Bill Castle-Roche, unidentified newspaper, in the possession of Susie Castle-Roche.

28. Information kindly provided by Glennys Gorman, daughter of Ken Saunders.

29. Most of the information about Bill Castle-Roche kindly provided by his daughter, Susie Castle-Roche, and his nephew, Tim Phillips.

30. *Cootamundra Herald*, 31 July 1946.

31. In the possession of Susie Castle-Roche.

32. *The Sydney Morning Herald*, 22 March 1963.

CHAPTER ELEVEN

1. I am indebted to both Matthew Crozier and Malcolm Peake for providing all information on modern Cavan.

2. Information on the ecology of Cavan and current ecological projects kindly provided by Alison Elvin of Natural Capital Pty Ltd.

3. Information kindly provided by Helen Crowley, Head of Sustainable Sourcing Information, Kering Group.

4. Information on the current state of the Australian wool industry kindly provided by Stuart McCulloch, CEO, Australian Wool Innovation.

5. Thomas Mitchell, *Three expeditions into the interior of eastern Australia: with descriptions of the recently explored region of Australia Felix and the present colony of New South Wales*, 1965, p 315.

PICTURE CREDITS

ACKNOWLEDGMENTS

My sincere thanks, firstly, to Alasdair and Prue Macleod, who conceived the idea of this book and invited me to write it. Secondly, I owe a great debt of assistance to Tim Phillips; his late mother, Lorna; and his cousin Susie Castle-Roche – all descendants of the Castle family, early pioneers at Cavan. Thirdly, I would like to thank my hard-working assistant, Val Wilkinson, for her thorough and inspired research. Fourthly, my thanks go to the eminent Geoffrey Blainey for endorsing this publication by way of a Foreword; and, fifthly, to the following people, all of whom contributed in some way to the content of this book: Rupert Murdoch, Matthew Crozier, Malcolm Peake, Tom Deery, Alison Elvin, Gavin Young, Carole Riley, Ian Cathles, Wal Merriman, Stuart McCullough, Edwina Winter, Cameron Archer, Judith Kelly, Chris Wood, Judy Sime, Rosslyn Waters, Patricia Clarke and Gloria Carlos, together with all staff who were so helpful at the Yass Archives, the Noel Butlin Archives at the ANU, the Mitchell Library, Sydney and the National Library of Australia, Canberra. Finally, I would like to thank Jane and Dominique de Stoop for many happy stays at Miragunyah during the course of my research, and the excellent editing and design teams at HarperCollins.

HarperCollins*Publishers*

First published in Australia in 2019
by HarperCollins*Publishers* Australia Pty Limited
ABN 36 009 913 517
harpercollins.com.au

HarperCollins*Publishers*
Level 13, 201 Elizabeth Street, Sydney NSW 2000, Australia
Unit D1, 63 Apollo Drive, Rosedale, Auckland 0632, New Zealand
A 53, Sector 57, Noida, UP, India
1 London Bridge Street, London, SE1 9GF, United Kingdom
Bay Adelaide Centre, East Tower, 22 Adelaide Street West, 41st floor, Toronto,
Ontario M5H 4E3, Canada
195 Broadway, New York NY 10007, USA

A catalogue record for this book is available from the National Library of Australia.

ISBN 978 1 4607 5742 0

Cover and internal design by Murray Batten
Cover image: courtesy of Lorna and Tim Phillips
Author image: Vanessa Taylor Photography
Colour reproduction by Graphic Print Group, South Australia
Printed and bound in China by RR Donnelley on 128gsm matt art

8 7 6 5 4 3 2 1   19 20 21 22 23

— *View over Bloomfield today*

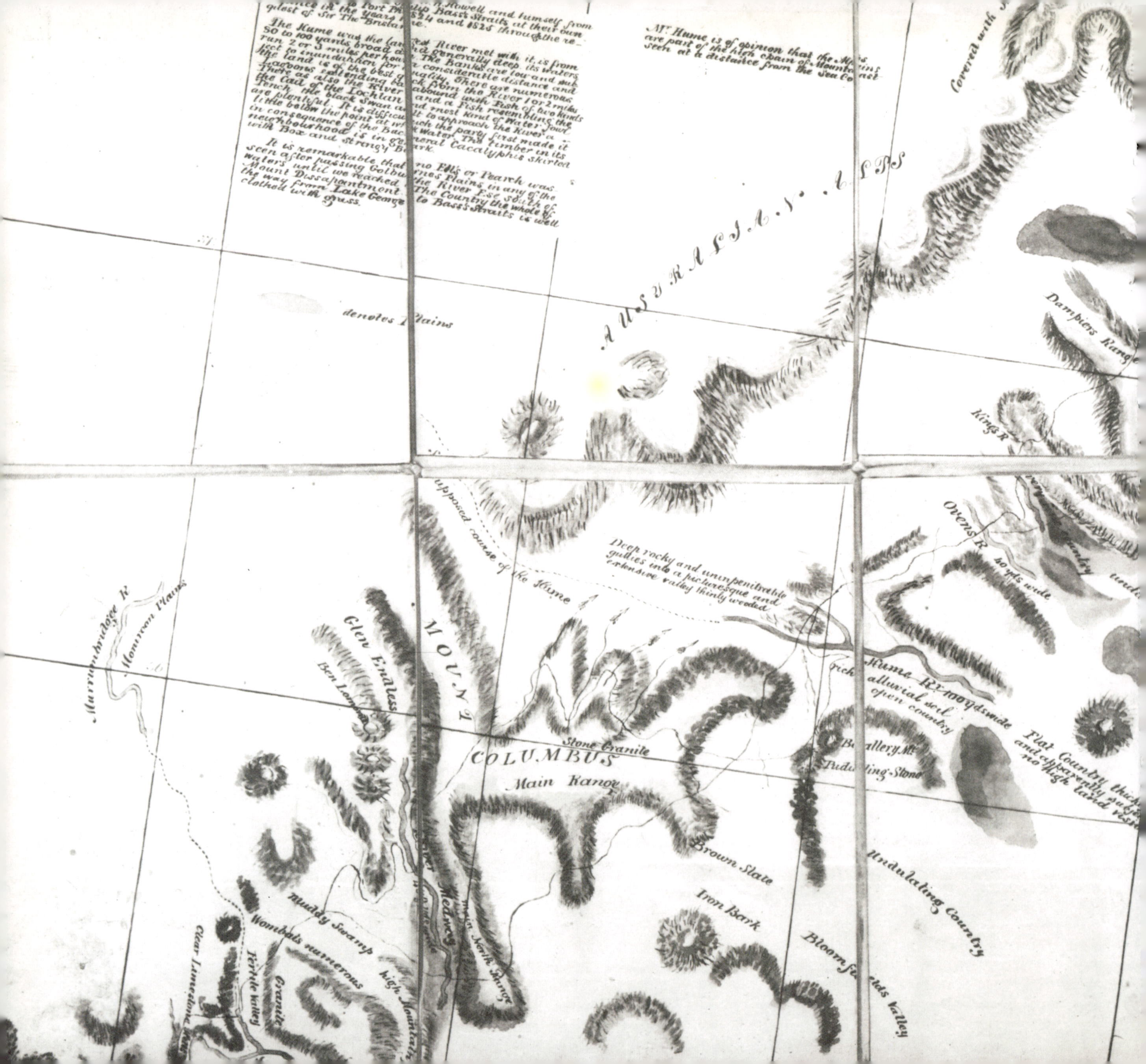

This is a historical hand-drawn map of southeastern Australia with annotations.

The Hume was the largest River met with, it is from 50 to 100 yards broad and generally deep, its waters run 2 or 3 miles per hour. The Banks are low and subject to inundation for a considerable distance and the land is of the best quality, there are numerous Lagoons extending back from the River 1 or 2 miles, there as also the Lochlun and Fish resembling the French, the Black Swan and most kind of Water Fowl are plentiful. It is difficult to approach the River, our little below the point at which the party first made it in consequence of the Back Water. The Neighbourhood is in general clothed with Box and Stringy Bark.

It is remarkable that no Falls or Rapids was seen after passing Golburnes Plains in any of the waters, until we reached the River Ese south of Mount Dissapointment. The Country the whole of the way from Lake George to Bass Straits is well clothed with Grass.

denotes Plains

Mr Hume is of opinion that the Alps are part of the high chain of Mountains seen at a distance from the Sea Coast

AUSTRALIAN ALPS

Dampiers Range

Kings R.

Ovens R. 60 yds wide

supposed course of the Hume

Deep rocky and unpenetrable gullies into a picturesque and extensive valley thinly wooded

Hume R. 100 yds wide alluvial soil open country

Murrumbridge R.

Monroom Plains

Glen Endless

Ben Lomond

MOUNT

COLUMBUS

Stone Granite

Main Range

Flat Country thick and appearently such no high land rests

B. Gallery M.

Puddling Stone

Brown Slate

Iron Bark

Bloom fields valley

Undulating Country

Muddy Swamp

Wombats numerous

high Mountains

Granite

Kettle Valley

Clear Limestone

Clear Limestone